Reevaluation of the Bedford-Berea Sequence in Ohio and Adjacent States: Forced Regression in a Foreland Basin

Jack C. Pashin
Geological Survey of Alabama
P.O. Box O
Tuscaloosa, Alabama 35486-9780

and

Frank R. Ettensohn
Department of Geological Sciences
University of Kentucky
Lexington, Kentucky 40506

SPECIAL PAPER

298

1995

Published by The Geological Society of America, Inc.
3300 Penrose Place, P.O. Box 9140, Boulder, Colorado 80301

Printed in U.S.A.

GSA Books Science Editor Richard A. Hoppin

Library of Congress Cataloging-in-Publication Data
Pashin, Jack C.
    Reevaluation of the Bedford-Berea sequence in Ohio and adjacent states : forced regression in a foreland basin / Jack C. Pashin and Frank R. Ettensohn.
        p.   cm. -- (Special paper ; 298)
    Includes bibliographical references.
    ISBN 0-8137-2298-5
    1. Sedimentation and deposition--Middle West.   2. Sedimentary basins--Middle West.   3. Geology, Stratigraphic--Devonian.
    I. Ettensohn, Frank R.   II. Title.   III. Title: Bedford-Berea sequence.   IV. Series: Special papers (Geological Society of America) ; 298.
    QE571.P365   1995
    551.4'45'0977--dc20

                                                                                      95-3060
                                                                                      CIP

Cover: The Bedford-Berea sequence in northeastern Ohio contains many large-scale sedimentary structures that are exposed in deep, scenic gorges. Shown on the cover is a giant crossbed, which crops out at Stebbins Gulch in Lake County. Similar crossbeds have been identified in other gorges and are interpreted to have formed in shallow marine environments during a delta-destructive phase of Bedford-Berea sedimentation.

10 9 8 7 6 5 4 3 2 1

Contents

iv   *Contents*

Preface

We prefer to think it was only yesterday, but our collaboration on the origin of the Bedford-Berea sequence began more than 12 years ago. Frank's interest in the Bedford Shale and Berea Sandstone goes back even further to the mid-1970s when he began analyzing Devonian and Mississippian black shale in the Appalachian basin. Most workers at the time treated the Bedford-Berea sequence as largely independent of the black shale but, by 1980, Frank and a graduate student, Tim Elam, showed that the Bedford-Berea was not only part of the black shale sequence but actually became black shale farther south in Kentucky. Frank's interest was stimulated even further in 1981, when he participated in a field trip led by Alan H. Coogan and Thomas L. Lewis, who showed him the exceptionally large and complex soft-sediment deformation structures that brought previous interpretations into question.

At the time, Frank did not know Jack, but Jack had been running wild across the Bedford-Berea type area near Cleveland since childhood collecting fossils and rocks. After completing his bachelor's degree in geology at Bradley University in December 1982, Jack had much of the basic knowledge he needed to understand what he had been seeing in and around Cleveland since his youth. Moreover, between his graduation from Bradley and his entry into graduate school in August 1983, Jack had plenty of time to consider the geology of the Cleveland area. He spent this time at the Shaker Lakes Regional Nature Center investigating Bedford-Berea strata exposed in the scenic gorges and waterfalls east of town. Little did he know of the profound influence these investigations would have on his future and, for this, Rich Horton and Dan Best of the nature center deserve special mention. The classic work of Pepper et al. (1954) was an excellent guide that gave Jack a major head start on understanding the stratigraphy and sedimentology of the Bedford-Berea sequence, and numerous lengthy conversations with Thomas L. Lewis and Joseph T. Hannibal gave him a glimpse of how many questions remained to be answered.

With so many questions still unanswered, Jack marched off to the University of Kentucky to begin graduate school, and Frank proceeded to put his nose to the grindstone. During the next three years, we both learned much about the Bedford-Berea sequence, and our first resulting paper characterized the little-known shelf-to-basin transition in the Bedford-Berea of eastern Kentucky and south-central Ohio and elaborated on some of the original ideas on epeiric sedimentation put forth by John L. Rich in the early 1950s (Pashin and Ettensohn, 1987). Next, we analyzed the fossil assemblages in the Bedford Shale to assess the ecologic dynamics of the dysaerobic zone (Pashin and Ettensohn, 1992a). The present report is a large-scale follow-up that integrates all our previous work with that of others to develop a regional account of the Bedford-Berea sequence in Ohio and the adjoining states. Our objective was to resolve the many controversies that have arisen since the classic study of Bedford-Berea sedimentation and paleogeography by Pepper et al. (1954), which was performed prior to the advent of plate tectonics and sequence stratigraphy.

This report is a synopsis of Jack's dissertation research, which was conducted at the University of Kentucky from 1985 to 1990. Funding for this research was provided by the Appalachian Basin Industrial Associates. Mark T. Baranoski, John M. Dennison, John C. Ferm, Thomas W. Kammer, Roy C. Kepferle, Ernest A. Mancini, Bruce R. Moore, Stephen O. Moshier, Nicholas Rast, and Robert C. Schumaker were exceptional sources of enlightenment and support during this investigation. Stephen F. Greb drafted the paleogeographic reconstructions, without which this report would have fallen far short.

Special thanks go to reviewers Thomas L. Lewis, Wallace de Witt, Jr., and Donald L. Woodrow, whose generous efforts and scholarly insight substantially improved the final manuscript. Don Woodrow's knowledge of Devonian stratigraphy helped crystallize our regional perspective, and Tom Lewis provided countless details that testify to his thorough and immediate knowledge of the Bedford-Berea sequence in northern Ohio. Lastly, Wallace de Witt's sage comments and gracious support of this publication are deeply appreciated, and we will always treasure his insights on the original studies of Rich and Pepper and colleagues.

Jack C. Pashin
Frank R. Ettensohn

Geological Society of America
Special Paper 298
1995

Reevaluation of the Bedford-Berea Sequence in Ohio and Adjacent States: Forced Regression in a Foreland Basin

ABSTRACT

The Upper Devonian Bedford-Berea sequence provided an early basis for models of epeiric sedimentation, but controversy regarding the origin of the sequence has arisen in recent years. This study utilized outcrop and subsurface data to help resolve this controversy and to identify factors that control depositional architecture in foreland basins. The Bedford-Berea is a siliciclastic succession that was deposited in the Appalachian foreland basin during a relaxational phase of the Acadian orogeny. The sequence represents a spectrum of depositional systems ranging from alluvial valleys to an oxygen-deficient basin floor and formed in response to a major forced regression that separated deposition of the Catskill and Pocono clastic wedges.

Reevaluation of the Bedford-Berea sequence demonstrates that the depositional architecture and paleogeographic history of foreland basins are much more elaborate than is commonly recognized. Tectonism, relict topography, differential compaction, and relative sea-level variation functioned collectively to determine the complex depositional history and paleogeography of the Bedford-Berea sequence. Among the salient features of Bedford-Berea paleogeography are an eastern platform and a western basin. The platform was characterized largely by erosion of Catskill sediment and subsequent deposition of aggradational valley-fill sequences, whereas the basin was characterized mainly by progradational delta and shelf deposits that overlie conformably the distalmost part of the Catskill clastic wedge. The platform and basin were differentiated by topography inherited from Catskill deposition, compaction of organic-rich sediment, and reactivation of basement structures along the Catskill shelf margin.

Bedford-Berea depositional history is divided into two episodes: basin filling and delta destruction. Basin filling was characterized by regressive fluvial systems that eroded Catskill strata and supplied prograding deltaic and shelf sediment to the western basin in the form of a lowstand wedge. Delta destruction began after the basin was filled with sediment. At this time, the basin was apparently undergoing flexural relaxation, which modified the lowstand wedge and gave rise to unusual facies patterns. Delta-front deposits in the western basin were uplifted and reworked, and a shelf silt blanket prograded back onto the rapidly subsiding eastern platform where estuaries were forming in the incised valleys.

INTRODUCTION

Few lithogenetic sequences have been as influential in shaping ideas of epeiric (epicontinental) sedimentation as the siliciclastic Bedford-Berea sequence of the Appalachian foreland basin. Rich (1951a,b) included the Bedford-Berea in the original model that served as the basis for subsequent accounts of epeiric sedimentation (van Siclen, 1958; Asquith, 1970; Heckel, 1972; Woodrow and Isley, 1983) and identified the clinoform, a foundation of modern sequence stratigraphy (Mitchum et al., 1977; Posamentier and Vail, 1988). Shortly after Rich's contributions, Pepper et al. (1954) produced a landmark study of Bedford-Berea sedimentation that presented fundamental ideas regarding regional paleogeography and is still

Pashin, J. C., and Ettensohn, F. R., 1995, Reevaluation of the Bedford-Berea Sequence in Ohio and Adjacent States: Forced Regression in a Foreland Basin: Boulder, Colorado, Geological Society of America Special Paper 298.

cited as a paradigm of epicontinental deltaic sedimentation (Krumbein and Sloss, 1963; Wanless et al., 1970; Frazier and Schwimmer, 1987; Stanley, 1989).

Recently, Pashin and Ettensohn (1987) applied an updated version of Rich's epeiric model (Woodrow and Isley, 1983) together with continental margin sedimentary models to the Bedford-Berea sequence of northeastern Kentucky and south-central Ohio. A major implication of Pashin and Ettensohn's study is that traditional epeiric sedimentary models, although widely applicable, are too simplistic to fully characterize epeiric sedimentation, especially where synsedimentary tectonism can be demonstrated. The investigations of Lewis (1968, 1976, 1988), Coogan et al. (1981), and Wells et al. (1991) have offered additional perspectives on the origin of the Bedford-Berea sequence and have challenged the paleogeographic framework put forth by Pepper et al. (1954).

The classic studies of Rich (1951a,b) and Pepper et al. (1954) were performed without the benefit of plate tectonics and sequence stratigraphy. Therefore, to better understand factors that control epeiric sedimentation, a major goal of this study was to reevaluate Bedford-Berea sedimentation and paleo-geography in Ohio and adjacent states in light of these profound geologic advances. In so doing, the objectives were to demonstrate the elaborately interwoven sedimentologic and tectonic variables that determined the extreme architectural complexity of the Bedford-Berea sequence, and to provide insight into the ways these variables interact in foreland basins.

Previous work

Studies of Bedford-Berea strata began in the early 19th century with the earliest geologic investigations of the eastern United States (Hildreth, 1836). Since then, research has been guided by scientific and technological advances, as well as historical events, including the growth of the oil industry. This discussion is a brief overview of the history of Bedford-Berea research; a thorough review is included in Pashin (1990).

An early understanding of the geology of North America was achieved in the 1830s, and it was during this time that the First Geological Survey of Ohio provided the first formal descriptions of the lithology and stratigraphy of what is now called the Bedford-Berea sequence (e.g., Briggs, 1838; Whittlesey, 1838). The First Survey was short-lived, however, and geologic endeavor effectively ceased until after the Civil War. After the war, a revitalized survey developed a geologic framework for the state that has, in many cases, remained intact. Newberry (1870) named the Bedford Shale and Berea Sandstone for exposures in northern Ohio (Fig. 1), and descriptive outcrop investigations continued until after World War II (e.g., Prosser, 1912; Cushing et al., 1931; Hyde, 1953).

Near the turn of the century, Bedford-Berea strata were recognized and described in neighboring states. White (1881) named the Cussewago and Corry sandstones for exposures in northwestern Pennsylvania (Fig. 1). Morse and Foerste (1909a,b, 1912) recognized the Bedford Shale and Berea Sand-

stone in eastern Kentucky, and Rominger (1876) recognized these units in the shallow subsurface of Michigan. Early in the 20th century, the first sedimentologic studies of the Bedford-Berea sequence were conducted. Most notable were classic studies of wave ripples (Hyde, 1911; Kindle, 1917; Bucher, 1919). Hyde, for example, was first to measure and map wave-ripple orientation on a regional basis. Soft-sediment deformation structures also were characterized in these early sedimentologic studies (Hyde, 1953; Cooper, 1943). By contrast, paleontologic investigation of the Bedford Shale began in the 1880s (Herrick, 1888) and is reviewed in Pashin and Ettensohn (1992a).

The birth and growth of the petroleum industry fostered a new understanding of Bedford-Berea strata in Ohio (Orton, 1879, 1888, 1893), Pennsylvania (Carll, 1890), West Virginia (Wasson and Wasson, 1929), Kentucky (Jillson, 1919), and Michigan (Newcombe, 1933). Orton (1879, 1888, 1893) stressed the lateral continuity of the Berea in eastern Ohio (Fig. 2) and realized that it extends into at least four other states, i.e., Pennsylvania, West Virginia, Kentucky, and Michigan. He also correlated the Berea with the Murrysville gas sand (Cussewago Sandstone) of Pennsylvania. Bownocker (1906) indicated that the Berea Sandstone was first drilled in 1860 or 1861. He also described the "Berea stray gas sand" (Second Berea Sand), which is the Cussewago–Second Berea siltstone belt of this study (Fig. 3). In West Virginia, Wasson and Wasson (1929) discovered that the most productive Berea reservoirs are in two linear sandstone bodies in the central part of the state called the Gay-Fink and Cabin Creek trends (Figs. 1, 2).

World War II necessitated efficient exploitation of America's petroleum resources and, as part of this effort, the U.S. Geological Survey initiated an 11-yr mapping program that focused on the Bedford-Berea sequence (Cohee and Underwood, 1944; Pepper et al., 1946; Demarest, 1946; de Witt, 1946, 1951; Rittenhouse, 1946). This project utilized a wealth of subsurface, outcrop, and petrographic data, and culminated in the landmark publication of Pepper et al. (1954), which synthesized the data into an unusually detailed regional investigation; the many maps in the work are still useful. Pepper et al. (1954) identified numerous depositional environments including fluvial, deltaic, beach-barrier, and marine shelf, and produced paleogeographic reconstructions that are considered the best for the time and are still included in textbooks (e.g., Stanley, 1989).

Rich (1951a,b) demonstrated the subdued nature of the epeiric sea floor in relation to that along continental margins. He further recognized the utility of clinoformal markers for characterizing basin stratigraphy and evolution. Rich's studies were exceptional contributions to the study of epeiric seas and offered a new perspective of Bedford-Berea sedimentation that favored a deep-water origin for black fissile shale like the Ohio Shale.

Especially important have been new developments in the understanding of petrology, sedimentology, paleontology, and tectonics, which have spurred on a multitude of studies throughout the distribution of the Bedford-Berea sequence.

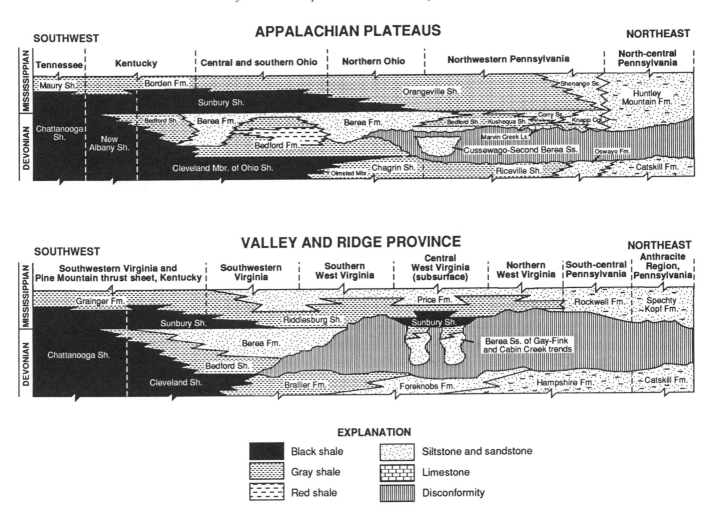

Figure 1. Stratigraphy of the Bedford-Berea sequence and bounding stratigraphic units in Ohio and adjacent states. The Bedford Berea sequence comprises the Bedford Shale, Berea Sandstone, Cussewago–Second Berea Sandstone, and equivalent units in the Appalachian foreland basin. The Bedford-Berea is bounded by black, fissile Cleveland and Sunbury shale and equivalent gray shale units throughout the region.

Included are studies from Ohio (Klein, 1967; Kohout and Malcuit, 1969; Lené and Owen, 1969; Rothman, 1978; Ferm, 1979; Coogan et al., 1981; Potter et al., 1983; Burrows, 1988; Duncan and Wells, 1992; Pashin and Ettensohn, 1992a), Pennsylvania (Schiner and Kimmel, 1972; Schiner and Gallaher, 1979), Kentucky (Morris and Pierce, 1967; Swager, 1979; Dillman, 1980; Elam, 1981; Ettensohn and Elam, 1985; Pashin, 1985; Pashin and Ettensohn, 1987, 1992b), West Virginia (Larese, 1974; Donaldson and Schumaker, 1981; Bjerstedt, 1986; Kammer and Bjerstedt, 1986; Boswell, 1988), Virginia (Kreisa and Bambach, 1973; Walls, 1975; Kepferle et al., 1981), and Michigan (McGregor, 1954; Asseez, 1969; Lilienthal, 1978; Ells, 1979; Fisher, 1980; Gutschick and Sandberg, 1991; Harrell et al., 1991; Matthews, 1993).

Most of these studies have operated within the framework of Pepper et al. (1954), and only the paleocurrent studies of Lewis (1968, 1976, 1988) and Lewis in Coogan et al. (1981)

have challenged the classic study. Whereas Pepper and other workers contended that the Berea of northern Ohio prograded southward from the Canadian Shield, Lewis demonstrated that crossbedding in the Berea is directed toward the north and west, suggesting source areas in the east or southeast. Lewis also used paleocurrent data to establish similar relationships in Bedford, Chagrin, and Cleveland shale, and he further indicated that sandstone bodies interpreted to be channel fills by Pepper et al. (1954) are actually the result of intense soft-sediment deformation. These ideas underscore the need to reevaluate the entire Bedford-Berea sequence, and it is to this end that the remainder of this report is committed.

Methods

Reevaluation of the Bedford-Berea sequence synthesized field, subsurface, and petrographic techniques. The field area is on the western and northern margins of the Appalachian basin in

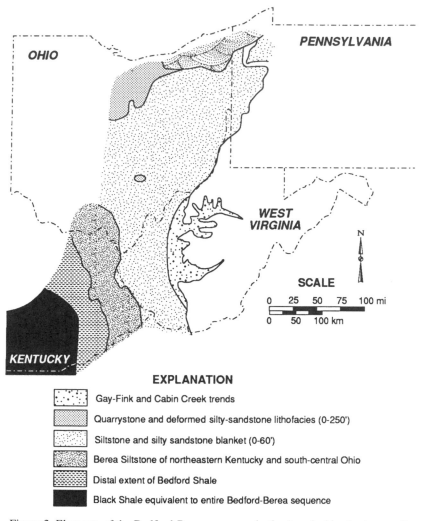

Figure 2. Elements of the Bedford-Berea sequence in the Appalachian basin: part I, Berea Sandstone.

Ohio and northwestern Pennsylvania (Fig. 4). Outcrops at 67 localities from Scioto County, Ohio, to Crawford County, Pennsylvania, were examined. Sections were measured using an alpha-numeric field classification, which is available in Pashin (1990), and is similar to a commonly used classification of coal-bearing rocks (Ferm and Weisenfluh, 1981). Directional data for paleocurrent analysis were collected as sections were measured. Azimuth, vector mean, vector magnitude, and standard deviation were computed using equations in Potter and Pettijohn (1977) and Krause and Geijer (1987). After outcrops were measured, sections were correlated lithostratigraphically, lithofacies were defined, and cross sections were made. Detailed measured sections and outcrop diagrams are in Pashin (1990).

Subsurface data, which include gamma-ray logs, cores, and cuttings from nearly 600 wells (Fig. 4) were analyzed at the Ohio Geological Survey. Gamma-ray logs are abundant in eastern Ohio, and the Bedford-Berea has a distinctive log signature, especially where bounded by radioactive black shale. One gamma-ray log per township was examined where possible,

and data for northwestern Pennsylvania were taken from Schiner and Kimmel (1972). Well logs, cores, and cuttings were correlated, and cross sections were made. The cross sections were then used to determine which stratigraphic intervals should be mapped to provide the most meaningful interpretation. Isopach maps of the most significant intervals were then constructed and contoured. In states adjacent to Ohio, many data are in works by Pepper et al. (1954), Larese (1974), Elam (1981), and Pashin (1985). These data form the basis for study outside Ohio, where interpretations are not as controversial. Hence, published data were simply integrated with new data from Ohio and northwestern Pennsylvania.

Thin sections of specimens from selected outcrops were prepared and described in terms of grain size, composition, and texture. Grain size was determined by measuring the long axes of 100 monocrystalline quartz grains per thin section. Composition was determined by counting a minimum of 300 points per thin section. Sandstone composition was plotted in the QFL triangle of Folk (1968) for classification and in the QFL and

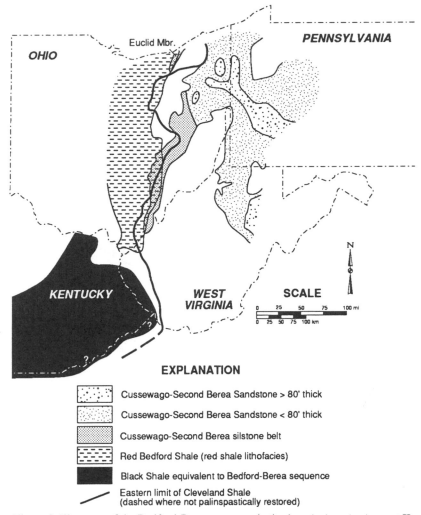

Figure 3. Elements of the Bedford-Berea sequence in the Appalachian basin: part II, Bedford Shale and Cussewago–Second Berea Sandstone.

Qm-F-Lt triangles of Dickinson et al. (1983) to determine provenance.

REGIONAL OVERVIEW

The Bedford-Berea sequence is part of a major succession of black fissile shale and intervening packages of light-colored shale, siltstone and sandstone, which extends across the North American craton (Fig. 5). The Bedford-Berea is a relatively thin sequence separating the much thicker Catskill (Upper Devonian) and Pocono (Lower Mississippian) clastic wedges, which were shed westward from the Acadian orogen (Ettensohn and Barron, 1981; Ettensohn, 1985a,b). According to Ettensohn (1985a, 1987), the Catskill and Pocono wedges prograded westward into an euxinic foreland basin that was formed flexurally as an Avalonian microplate converged obliquely with the Laurussian continent (Fig. 6).

Stratigraphic framework

Bedford Shale, Berea Sandstone, and Cussewago-Second Berea Sandstone. The Upper Devonian Bedford-Berea sequence is the stratigraphic interval between the black fissile Cleveland Shale and the Sunbury Shale or their equivalents in Ohio and adjacent states (Fig. 1). East of the pinchout of the Cleveland, the Bedford-Berea is underlain by the Chagrin Shale in Ohio and equivalent strata in the surrounding states (Pepper et al., 1954; de Witt et al., 1993). The Sunbury Shale overlies the Bedford-Berea in most of eastern Ohio and passes eastward into gray shale of the Orangeville and Cuyahoga Formations (Pepper et al., 1954).

The Bedford-Berea sequence comprises the Bedford Formation, the Berea Formation, and the Cussewago–Second Berea Sandstone (Fig. 1). The Bedford Formation, or Bedford Shale, extends throughout eastern Ohio, eastern Kentucky, and northwestern Pennsylvania; the shale may be present locally in

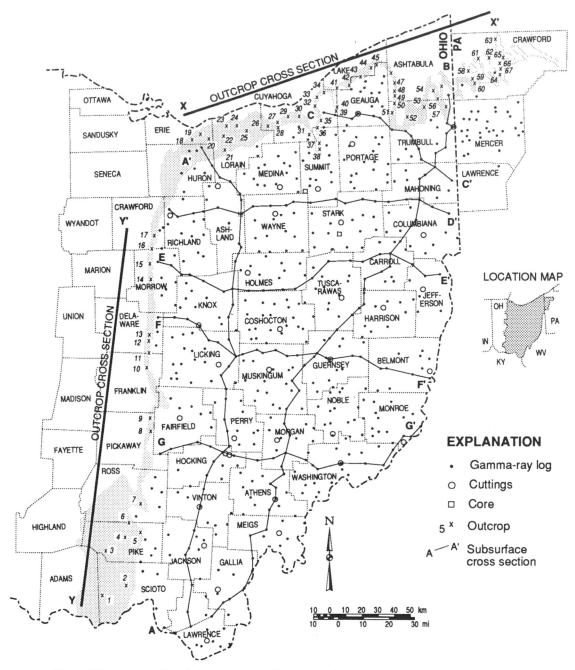

Figure 4. Index map of detailed study area in Ohio and northwestern Pennsylvania showing location of outcrops, wells, and lines of cross section.

westernmost West Virginia and has been recognized in Virginia (Kepferle et al., 1981) and in the Michigan basin of Michigan, northwestern Ohio, and northeastern Indiana (Pepper et al., 1954; Matthews, 1993). The Bedford overlies the Cleveland Shale, Chagrin Shale, and correlative units, and generally underlies the Berea Sandstone (Figs. 7–10). In northeasternmost Ohio and northwestern Pennsylvania, where stratigraphic relationships are extremely complex, the Bedford forms a tongue of shale within the Berea.

The Bedford Formation contains two major color phases (Fig. 1). Gray shale containing numerous laminae and thin beds of siltstone is the most widespread phase, whereas red or brown shale is restricted to the Appalachian Plateau and is present in a belt extending from north-central Ohio into northeasternmost Kentucky and westernmost West Virginia (Pepper et al., 1954) (Figs. 1, 3). In northeastern Ohio, the gray phase contains two major siltstone bodies called the Euclid and Sagamore Members (Prosser, 1912; de Witt, 1951) (Figs. 3, 7). The Second

Berea Sandstone of southeastern Ohio (Fig. 3) is lithologically similar to these members (Pepper et al., 1954) and, as discussed later, is an extension of the Cussewago Sandstone.

The Berea Formation is widespread in eastern Ohio, eastern Kentucky, northwestern Pennsylvania, western and central West Virginia, and southwestern Virginia (Pepper et al., 1954; de Witt et al., 1993) (Figs. 1, 2). The Berea is also present in the Michigan basin (Cohee and Underwood, 1944; Ells, 1979; Gutschick

and Sandberg, 1991). Sandstone is restricted to northern Ohio, West Virginia, eastern Michigan, and part of southwestern Virginia, whereas siltstone is present throughout the remainder of the region (Hyde, 1953; Ettensohn and Dever, 1979).

The Cussewago Sandstone, or Cussewago–Second Berea Sandstone, is present in northeastern Ohio, western Pennsylvania, and northeastern West Virginia (Figs. 1, 3). The Cussewago is widespread in the subsurface and crops out only in northeast-

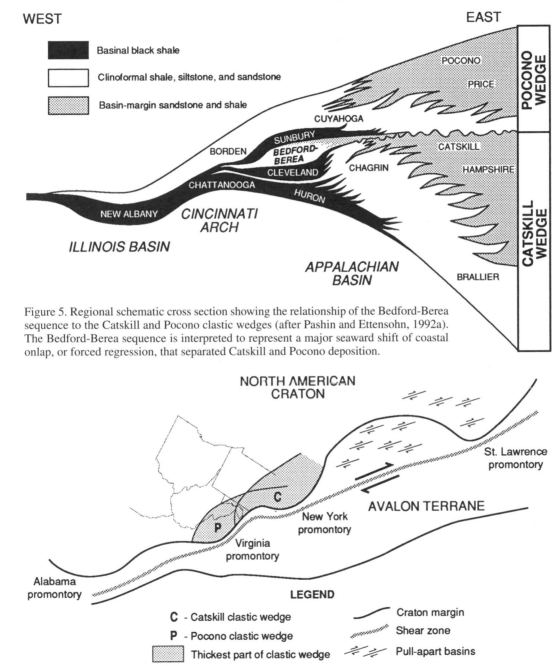

Figure 5. Regional schematic cross section showing the relationship of the Bedford-Berea sequence to the Catskill and Pocono clastic wedges (after Pashin and Ettensohn, 1992a). The Bedford-Berea sequence is interpreted to represent a major seaward shift of coastal onlap, or forced regression, that separated Catskill and Pocono deposition.

Figure 6. Tectonic setting of the Catskill and Pocono clastic wedges (after Ettensohn, 1985b). The Catskill and Pocono wedges were shed from the Appalachian orogen during the Acadian orogeny. The Catskill wedge was deposited after an Avalonian microplate collided with the New York continental promontory, whereas the younger Pocono wedge was deposited after collision with the Virginia promontory.

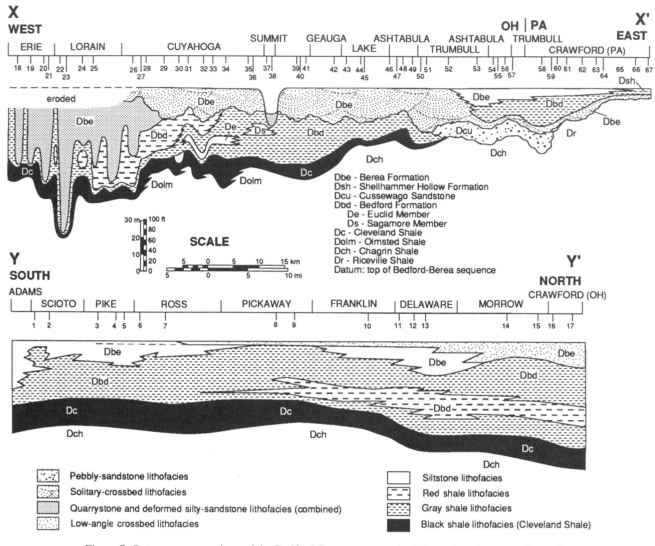

Figure 7. Outcrop cross sections of the Bedford-Berea sequence in Ohio and northwestern Pennsylvania. See Figure 4 for location.

ern Ohio and northwestern Pennsylvania. In outcrop the formation is composed largely of friable sandstone containing quartz pebbles. The sandstone underlies the Bedford and Berea Formations and overlies the Chagrin Shale and equivalent strata (Figs. 1, 7–9). In the subsurface of Pennsylvania, the Cussewago rests disconformably on the Catskill redbeds (Pepper et al., 1954). The Cussewago was for some time thought to be separate from the Berea (de Witt, 1951; Pepper et al., 1954) but, as shown in cross sections C-C′ (Fig. 9) and X-X′ (Fig. 7), the two units are connected. Additionally, cross section X-X′ indicates that the Berea Sandstone extends above the Cussewago Sandstone into northwestern Pennsylvania.

The Cussewago Sandstone is mainly east of the pinchout of the Cleveland Shale and overlies gray Chagrin and Riceville shale (Figs. 1, 7, 8). Gamma-ray logs establish that the Cussewago extends southward into the Second Berea Sand (Cussewago–Second Berea siltstone belt) (Figs. 3, 8), and a

Cussewago equivalent that was included in the Berea by Pepper et al. (1954) is identifiable in much of east-central Ohio. Therefore, the Cussewago Sandstone and the Second Berea Sandstone, which were previously thought to be separate (Pepper et al., 1954), are unified in a single unit called the Cussewago–Second Berea Sandstone (Fig. 1).

Tracing the Cussewago into the Second Berea is important for establishing stratigraphic relationships, because the western part of the Cussewago–Second Berea Sandstone overlies the Cleveland Shale in Tuscarawas, Guernsey, Muskingum, and Morgan Counties (Figs. 8–10). This relationship verifies that the Cussewago–Second Berea is younger than the Cleveland and justifies treatment of the Cussewago and equivalent strata as the base of the Bedford-Berea sequence where the Cleveland Shale is absent. Difficulty in correlation arises where both the Cussewago–Second Berea and the Cleveland are absent in southeastern Ohio. Here, gray Bedford Shale overlies gray

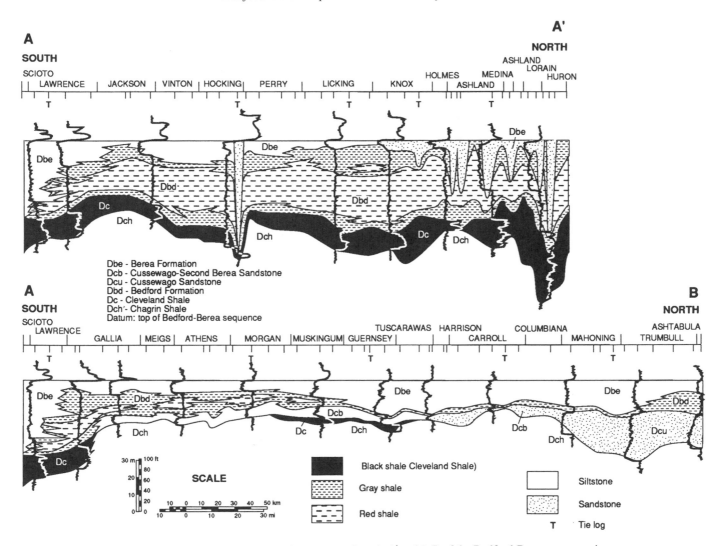

Figure 8. North-south gamma-ray log cross sections A-A' and A-B of the Bedford-Berea sequence in Ohio. See Figure 4 for location.

Chagrin Shale, and the contact cannot be identified in well logs or cuttings (Figs. 8–10). This difficulty was resolved, however, because isolated bodies of siltstone are at the approximate stratigraphic level of the Cussewago–Second Berea and thus provide some control for correlation (Fig. 10).

The Bedford-Berea sequence is commonly thicker than 45 m (150 ft) in the western part of the study area and is locally thicker than 66 m (200 ft) in north-central Ohio (Fig. 11). By contrast, the sequence is scarcely more than 23 m (75 ft) thick throughout most of southeastern Ohio and is locally thinner than 16 m (50 ft). The Bedford-Berea thins eastward by as much as 30 m (100 ft) across a north-south line extending from Coshocton County to Perry County. The line of thinning, which is delineated by the 100-ft (33-m) contour, represents a major depositional boundary separating an *eastern platform* from a *western basin* (Figs. 9–11). The platform margin was a salient feature of Bedford-Berea deposition that had a profound effect on facies distribution and, hence, paleogeography.

The platform margin can be traced southward into Gallia County but is not as clearly defined as in central Ohio. In northeastern Ohio, identifying the platform margin is complicated by absence of the Cleveland Shale and the Cussewago–Second Berea Sandstone. Even so, data from surrounding areas are abundant enough to show that the margin turns east and extends far into northeastern Ohio. Thickening of the Bedford-Berea sequence to more than 30 m (100 ft) in Stark, Mahoning, and Trumbull Counties is related to thickening of the Cussewago–Second Berea Sandstone (Figs. 11, 12).

Equivalent formations. Although this study is concerned mainly with the Bedford, Berea, and Cussewago Formations, discussion of equivalent strata and their bounding units in the Appalachian and Michigan basins is essential for regional perspective. In northwestern Pennsylvania, Bedford-like shale and siltstone called the Shellhammer Hollow Formation separates the Cussewago Sandstone and Orangeville Shale (Sunbury equivalent) in a belt approximately 16 km (10 mi) wide where

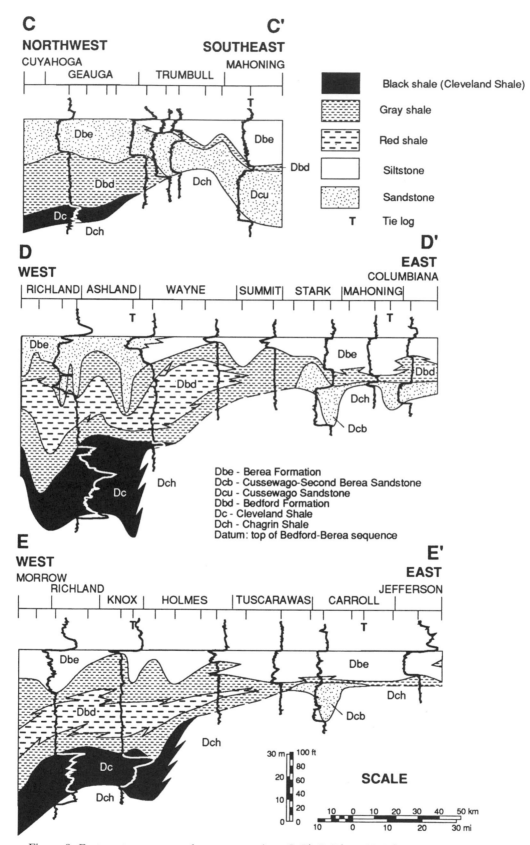

Figure 9. East-west gamma-ray log cross sections C-C′, D-D′, and E-E′ of the Bedford-Berea sequence in Ohio. See Figure 4 for location.

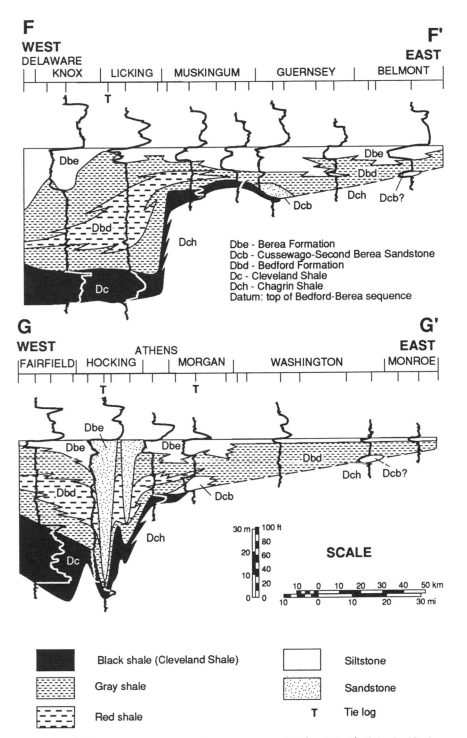

Figure 10. East-west gamma-ray log cross sections F-F′ and G-G′ of the Bedford-Berea sequence in Ohio. See Figure 4 for location.

the siltstone of the Berea Formation is absent (de Witt, 1951) (Fig. 1). Uppermost Chagrin equivalents in this same area are referred to as the Riceville Shale. Eastward, the Shellhammer Hollow Formation passes into shale similar to the Bedford and Riceville, whereas upper part of the Shellhammer Hollow passes into the Corry Sandstone, which is lithologically similar to and a probable correlative of the Berea. A few miles east of the margin of the Corry, the Cussewago pinches out.

Caster (1934) traced the Corry Sandstone into north-central Pennsylvania, where the unit overlies the so-called Knapp formational suite (Fig. 1). In this region, the Orangeville Shale is present, but the Riceville Shale passes eastward into the coarser Oswayo Formation. According to Caster, the Knapp contains three formations named, in ascending order, the Marvin Creek Limestone, the Kushequa Shale, and the Knapp Conglomerate. Brachiopod assemblages in the Knapp and Corry indicate a Late Devonian age (Sass, 1960; Carter and Kammer, 1990).

In north-central Pennsylvania, the Knapp passes eastward into the gray and red shale, siltstone, sandstone, and conglomerate of the lower part of the Huntley Mountain Formation, which also contains strata of Mississippian age (Berg and Edmunds, 1979; Berg et al., 1983) (Fig. 1). Whereas the Knapp Conglom-

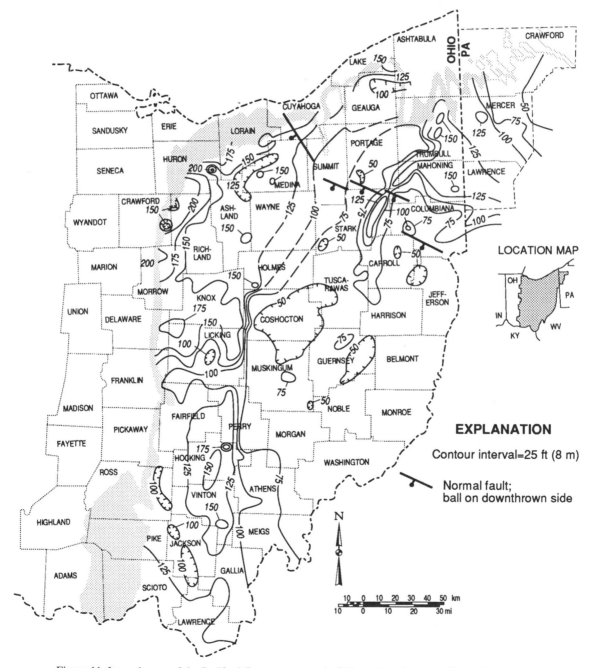

Figure 11. Isopach map of the Bedford-Berea sequence in Ohio and northwestern Pennsylvania.

erate contains marine fossils, the Huntley Mountain Formation contains plant fossils (Berg and Edmunds, 1979). Farther south in the Valley and Ridge province, strata equivalent to the Huntley Mountain Formation are called the Rockwell Formation (Berg and Edmunds, 1979; Berg et al., 1983). In the Anthracite Region, Bedford-Berea strata may be within the Spechty Kopf Formation, which resembles the Huntley Mountain Formation and contains diamictite units of enigmatic origin, as well as some thin coal beds (Berg et al., 1983; Sevon, 1985).

In western and central West Virginia, drillers recognize the

Sunbury Shale above the Berea, although black fissile shale is lacking in the remainder of the state (Pepper et al., 1954) (Fig. 1). The Berea is absent in the eastern part of the state (Fig. 2), but the Cussewago Sandstone may extend into the Valley and Ridge province of West Virginia and Pennsylvania, where it has apparently been reworked and incorporated into the base of the transgressive Riddlesburg Shale (Bjerstedt, 1986; Kammer and Bjerstedt, 1986; Bjerstedt and Kammer, 1988). In Virginia, transgressive sandstone and conglomerate units equivalent to the Bedford-Berea have also been recognized in the Cloyd

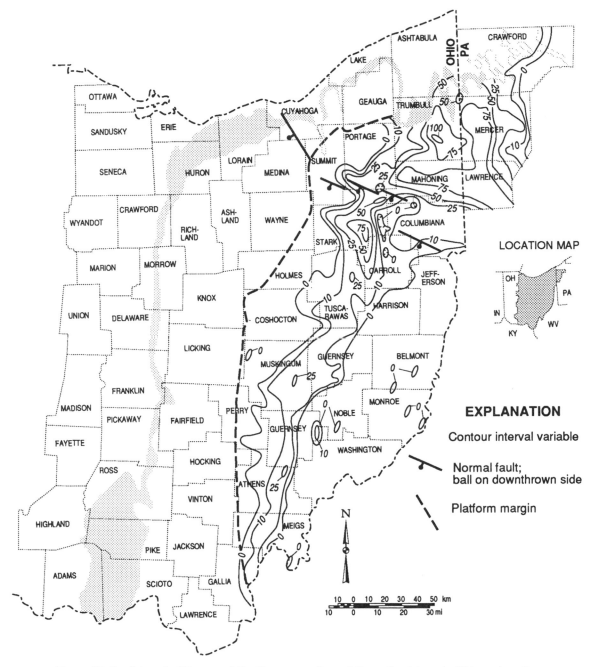

Figure 12. Sandstone isolith map of the Cussewago–Second Berea Sandstone in Ohio and northwestern Pennsylvania.

Conglomerate Member of the Price Formation (Kreisa and Bambach, 1973; Rossbach and Dennison, 1994).

The relationship between Bedford-Berea and adjacent strata in eastern Kentucky is similar to that in Ohio but, in the northeasternmost part of the state, the Berea is divided into upper and lower tongues; the lower tongue rests directly upon the Cleveland Shale, and the two tongues are separated by a northeastwardly thinning wedge of the Bedford Shale (Morris and Pierce, 1967; Pashin and Ettensohn, 1987; Ettensohn et al., 1988) (Fig. 1). In the part of east-central and southeastern Kentucky where the Berea is absent, Bedford and equivalent gray shale units separate the Sunbury and Cleveland or their equivalents, which are mapped with the New Albany and Chattanooga Formations (Ettensohn and Elam, 1985; Pashin and Ettensohn, 1987). Black shale equivalent to the Bedford-Berea has been recognized in much of eastern Kentucky (Swager, 1979; Elam, 1981; Ettensohn and Elam, 1985) (Fig. 3) and is included in the Ohio, New Albany, and Chattanooga Formations (Fig. 1); no such black shale has been identified in Ohio or West Virginia.

The Bedford Shale and Berea Sandstone also are present in eastern Michigan (Cohee and Underwood, 1944; Cohee, 1965). Equivalent gray dolomitic shale and localized dolostone bodies are present in the Ellsworth Shale of western Michigan (Hale, 1941). Shale and siltstone equivalent to the Bedford-Berea sequence also have been recognized in the Port Lambton Beds, which overlie black shale of the Kettle Point Formation and are restricted to a small area of westernmost Ontario (Sanford and Brady, 1955).

Structural framework

Evidence from the Bedford-Berea sequence of Kentucky suggests that basement structure had a major influence on epeiric sea-floor topography (Pashin and Ettensohn, 1987). Hence, explanation of Bedford-Berea sedimentation elsewhere in the Appalachian basin may be incomplete without testing relationships between rock distribution and regional structure. The Bedford-Berea sequence is widespread in the northwestern part of the Appalachian basin, an elongate area bounded by the Cincinnati and Findlay arches in the west, the Algonquin arch in the northwest, the Frontenac arch and the Adirondack dome in the north, and the Blue Ridge anticlinorium in the east. The eastern margin of the basin contains the numerous folds and thrust faults of the Valley and Ridge province and the Cumberland Plateau, whereas the western part contains normal basement faults of the Rome trough, cross-strike structures of enigmatic origin, and basement structures of the Grenville orogen. Open folds, including the Burning Springs anticline and the Cambridge arch, also are present in the central part of the basin (Fig. 13). These folds are shallow, detached structures formed during the Alleghanian orogeny and thus postdate Bedford-Berea deposition (Rodgers, 1963; Gray, 1982).

The Rome trough extends from Pennsylvania to eastern Kentucky (Fig. 13) and is part of a major set of basement faults formed during Late Precambrian to Cambrian Iapetan rifting (e.g., Thomas, 1991). In West Virginia, the Rome trough is a major graben, but in parts of Pennsylvania, the structure is composed mainly of en echelon faults that are downthrown toward the southeast (Harris, 1975; Wagner, 1976; Ammerman and Keller, 1979). In West Virginia, several basement faults strike oblique to the Rome trough and define an associated set of fault blocks. An uplifted area east of the Rome trough has variously been called the Pocono dome, the West Virginia positive area, and the West Virginia dome (Kammer and Bjerstedt, 1986). Donaldson and Schumaker (1981) indicated that the structure is the highest part of a south-dipping basement block. North of the uplift, Bjerstedt and Kammer (1988) identified a downdropped basement block that they named the "cross-over rift," and they called the uplifted area north of the cross-over rift the "Tri-State Block."

Two major cross-strike structures, the Tyrone–Mount Union lineament and the Transylvania fault, trend northwest across regional strike and are downthrown toward the southwest (Root and Hoskins, 1978; Rodgers and Anderson, 1984) (Fig. 13). Root and MacWilliams (1986) recognized that the Transylvania fault is contiguous with the Middleburg, Akron, Suffield, Smith Township, and Highlandtown faults of northeastern Ohio (Gray, 1982). Rodgers and Anderson (1984) suggested that the Tyrone–Mount Union lineament is a basement-fault system that was downthrown to the southwest during the late Paleozoic. The origin of the cross-strike structures is debatable, but Root and Hoskins (1978) cited offsets of gravity anomalies along the Transylvania fault as evidence for a Precambrian transform zone.

Samples from basement tests contain mainly granite gneiss and amphibolite, establishing that the Grenville basement is a metamorphic terrane (Lucius and von Froese, 1988). Characteristics of the terrane have until recently been inferred mainly from gravity and magnetic data. In southern and central Ohio, the Bedford-Berea platform margin follows a pronounced lineament on the gravity anomaly map (Fig. 14). In northern Ohio, the platform margin is less well defined and is east of a gravity high of unknown origin centered in Wayne County.

A COCORP seismic profile (Pratt et al., 1989) shows that Grenville basement in Knox County contains east-dipping reflectors that extend to the Moho and coincide with the position of the Grenville front tectonic zone of Lucius and von Froese (1988). In and east of Coshocton County, the basement contains similar, west-dipping reflectors that make up the so-called Coshocton zone and have been interpreted by Pratt et al. (1989) and Culotta et al. (1990) to represent a separate tectonic terrane. The boundary between the two sets of dipping reflectors coincides with the platform margin, suggesting a reactivated Grenville suture influenced Bedford-Berea sedimentation in central Ohio. The suture also appears to correspond with the so-called Newark arch, which may have influenced lower Paleozoic sedimentation (Dolly and Busch, 1972).

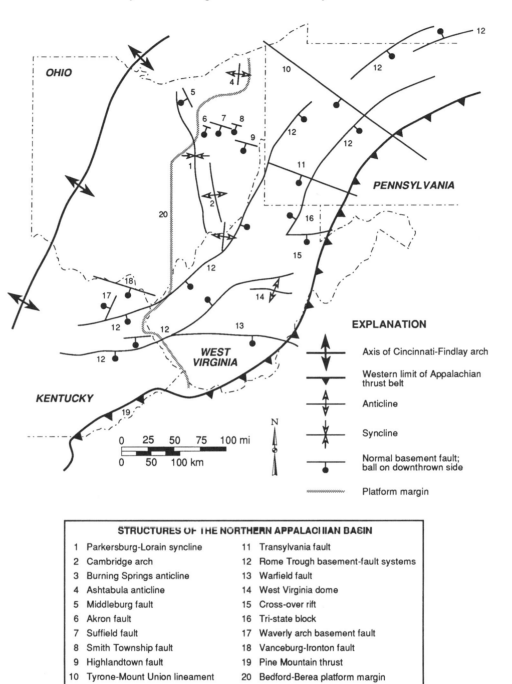

Figure 13. Major tectonic structures in the northern part of the Appalachian basin. Structures affecting Bedford-Berea deposition were formed primarily during Late Precambrian–Early Cambrian rifting. Older Grenvillian structures may also have been effective locally.

EXPLANATION

Axis of Cincinnati-Findlay arch

Western limit of Appalachian thrust belt

Anticline

Syncline

Normal basement fault; ball on downthrown side

Platform margin

STRUCTURES OF THE NORTHERN APPALACHIAN BASIN	
1 Parkersburg-Lorain syncline	11 Transylvania fault
2 Cambridge arch	12 Rome Trough basement-fault systems
3 Burning Springs anticline	13 Warfield fault
4 Ashtabula anticline	14 West Virginia dome
5 Middleburg fault	15 Cross-over rift
6 Akron fault	16 Tri-state block
7 Suffield fault	17 Waverly arch basement fault
8 Smith Township fault	18 Vanceburg-Ironton fault
9 Highlandtown fault	19 Pine Mountain thrust
10 Tyrone-Mount Union lineament	20 Bedford-Berea platform margin

LITHOFACIES ANALYSIS

This chapter describes the basic rock types of the Bedford-Berea sequence and defines nine lithofacies based on assemblages of those rock types and their sedimentary structures. Brief descriptions of the basic rock types, which are sandstone, siltstone, and shale, are given in the following paragraphs. More thorough descriptions of lithology, geophysical log characteristics, and petrology are in Pashin (1990). The sections following the basic lithologic descriptions define, describe, and interpret each lithofacies of the Bedford-Berea sequence.

Three types of sandstone were recognized: pebbly sandstone, sandstone, and silty sandstone. Pebbly sandstone is yellowish-brown, coarse grained, and is generally friable; it contains granules, pebbles, and cobbles of quartz, sideritic siltstone, and shale. Sandstone is typically yellowish-brown, very

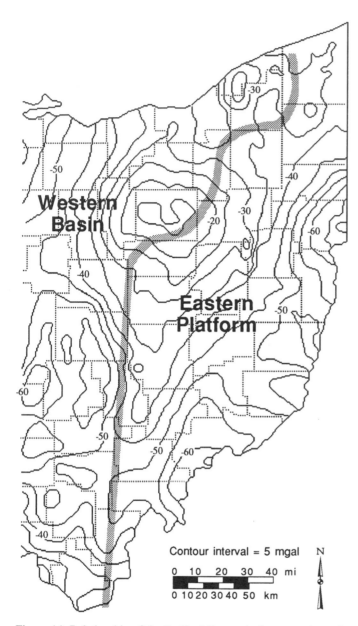

Figure 14. Relationship of the Bedford-Berea platform margin to the gravity anomaly map of Lucius and von Froese (1988).

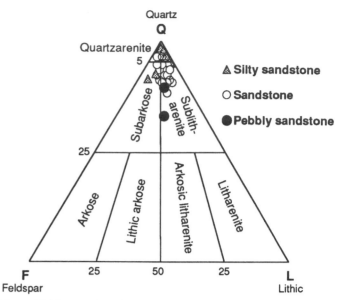

Figure 15. Ternary plot of sandstone composition in the Bedford-Berea sequence. Silty sandstone is mainly quartzarenite, and sandstone and pebbly sandstone range in composition from quartzarenite to sublitharenite and subarkose.

Siltstone is one of the most abundant rock types in the Bedford-Berea sequence and, without exception, is well cemented and is resistant to weathering. The rock is composed mainly of monocrystalline quartz with a lesser amount of mica, clay-mica matrix, and intergranular ferroan calcite cement. Coarse-grained silt predominates in medium- and thick-bedded siltstone, but the laminated and thin-bedded siltstone is more variable and contains coarse- to fine-grained silt. The siltstone is cemented primarily with quartz with minor amounts of intergranular ferroan calcite and framboidal marcasite.

Three types of shale were identified in the Bedford-Berea sequence: gray shale, red shale, and black shale. Gray shale is silty and typically has an irregular, platy fissility and in places contains siderite concretions and marcasite nodules. Red shale is much richer in clay than gray shale, is softer, and typically has a well-developed, platy fissility; the shale characteristically lacks concretions and marcasite. The mineralogy of the red and gray shale is detailed in Lamborn et al. (1938). Black shale is distinctive because of color, brittleness, fissility, and radioactivity; it is restricted to the Cleveland Shale and the Sunbury Shale in Ohio. The shale is silty and locally contains nodules of marcasite and, less commonly, phosphate. Black shale owes its color to high kerogen content, which is mainly oil-prone matrix bituminite and alginite (*Tasmanites*).

Pebbly-sandstone lithofacies

Characteristics. The pebbly-sandstone lithofacies is composed almost exclusively of thick-bedded pebbly sandstone and is restricted in outcrop to the Cussewago Sandstone in north-

fine to medium grained, and moderately friable. By contrast, silty sandstone is generally very fine grained, light medium gray, and tightly cemented. Pebbly sandstone is the most poorly sorted and mineralogically immature rock type in the Bedford-Berea sequence. Framework grains include monocrystalline quartz, polycrystalline quartz, feldspar, chert, and chlorite schist. The sandstone ranges in composition from quartzarenite to sublitharenite and subarkose (Fig. 15). Silty sandstone is generally quartzarenite, whereas sandstone is variable in composition, and pebbly sandstone is litharenite. Quartz cement predominates, but in places the sandstone is cemented with ferroan calcite, ferroan dolomite, or marcasite.

eastern Ohio and northwestern Pennsylvania (Fig. 7). The contacts of the lithofacies are not exposed, but the pebbly-sandstone lithofacies is absent in the easternmost part of the outcrop belt where the Berea Sandstone overlies the Riceville Shale (Figs. 1, 7).

The pebbly sandstone contains abundant trough and planar crossbeds; quartz granules and pebbles are scattered throughout the sandstone and are locally concentrated in lenses along foresets. Only three reliable crossbed azimuths were available,

but despite few data, the readings range from 313° to 364° (Fig. 16) and have a combined vector magnitude of 0.99.

Although this lithofacies has a limited outcrop distribution that shows few facies relationships, subsurface data are revealing. Pebbly sandstone is particularly abundant in the Cussewago–Second Berea Sandstone, the Gay-Fink trend, and the Cabin Creek trend (Figs. 2, 3), which have all been described from the subsurface. Subsurface characterization establishes that the pebbly-sandstone lithofacies is present in a series of linear bodies

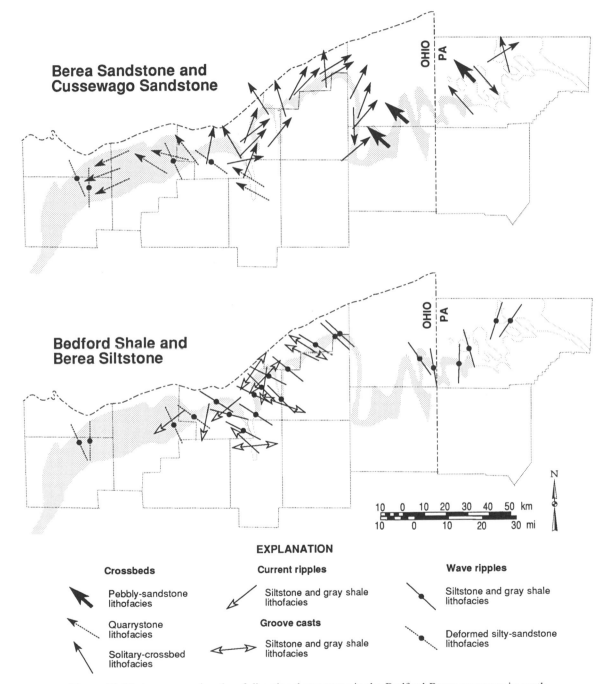

Figure 16. Vector-mean azimuths of directional structures in the Bedford-Berea sequence in northeastern Ohio and northwestern Pennsylvania.

that are the principal coarse-grained sediment axes of the Bedford-Berea sequence. These bodies are described and interpreted below.

Cussewago–Second Berea Sandstone. The Cussewago–Second Berea Sandstone varies in thickness from 23 to 33 m (75 to 100 ft) in parts of Mahoning and Trumbull Counties, northeastern Ohio, and Mercer and Lawrence Counties, northwestern Pennsylvania (Figs. 12, 17). In Lawrence County, Pepper et al. (1954) identified a linear axial sandstone trend that is oriented northwest and extends southeast of the map area. The

75-ft (23-m) contour in the isopach map (Fig. 12) shows that the trend bifurcates in northwestern Pennsylvania at an approximate angle of 90°. The western limb is mainly in Mahoning County and is directed due west. An extension of the limb is outlined by the 25-ft (8-m) contour and closely parallels the Suffield fault in Portage County. In Crawford County, the other limb is directed due north and is broader, thinning abruptly toward the east and gradually toward the west. Centered between the two limbs in Trumbull County, Ohio, is an isolated body of thick sandstone defined by the 75-ft (22-m) contour.

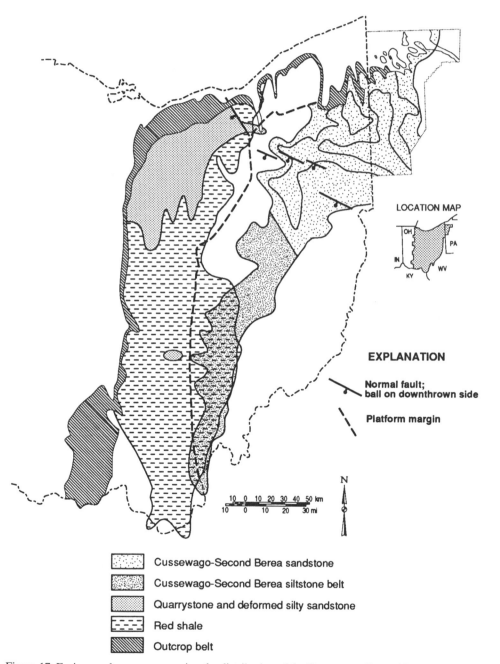

Figure 17. Facies overlay map comparing the distribution of the Cussewago–Second Berea Sandstone, the quarrystone lithofacies of the Berea Sandstone, and the red shale lithofacies of the Bedford Shale.

These features are superimposed on a larger, lobate form outlined by the 50-ft (15-m) contour.

The lobe thins gradually toward the outcrop belt in the north, but the 25-ft (8-m) contour outlines a hooklike trend that curves southward from Mahoning County to Tuscarawas County. A minor sandstone axis thicker than 3 m (10 ft) extends from the main Cussewago axis into an embayment defined by the hooklike form in Columbiana and Carroll Counties (Fig.

12). Near the southern extremity of the hook in Stark and Carroll Counties, the sandstone locally exceeds 22 m (75 ft) in thickness. In a core from this area the Cussewago–Second Berea sharply overlies the gray shale and siltstone of the Chagrin instead of the Bedford. The core includes a thick bed of quartz-pebble conglomerate that grades upward into crossbedded sandstone and, in turn, into wavy bedded shale and silty sandstone with current-ripple cross-laminae (Fig. 18).

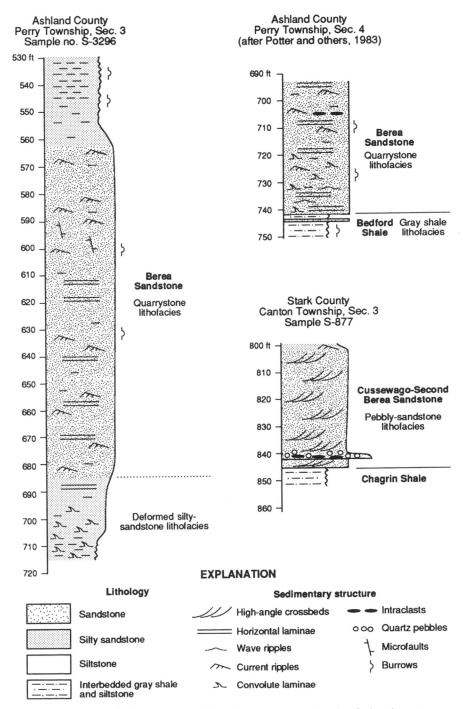

Figure 18. Selected cores of the Bedford-Berea sequence showing facies characteristics in the subsurface of Ohio.

Thick Cussewago–Second Berea Sandstone correlates with thickening of the Bedford-Berea sequence to more than 46 m (150 ft) on the eastern platform (Figs. 11, 12). The 125-ft (38-m) contour on the Bedford-Berea isopach map outlines a bifurcating trend in Mahoning and Trumbull Counties (Fig. 11); however, that bifurcation is in a location different from that in the Cussewago–Second Berea. Sandstone passes southward into siltstone in Tuscarawas County, and the Cussewago–Second Berea forms a sublinear belt that includes the Second Berea Sand of Pepper et al. (1954) (Figs. 12, 17). The Cussewago–Second Berea siltstone belt is included in the siltstone lithofacies, which is discussed in a later section.

Gay-Fink and Cabin Creek trends. Sandstone of the Gay-Fink and Cabin Creek trends (Figs. 19, 20) contains quartz pebbles at the base and tends to fine upward into shale (Larese, 1974); the Sunbury Shale extends farther east within the trends than in adjacent areas (Larese, 1974). Axes of the trends are defined by the 20-ft (6-m) contour (Fig. 19). The Gay-Fink trend branches and has a sublinear axis, whereas the Cabin

Creek trend is funnel shaped and has a sinuous axis; Bedford-Berea strata are absent between the two trends. The Gay-Fink trend is bounded by the faults of the Rome trough (Fig. 20). The Cabin Creek trend is southwest of the Rome trough, and the southwestern margin of the trend follows the Warfield fault. The thick axial sandstone of the Cabin Creek trend extends more than 10 km (7 mi) westward into a blanket of sandstone and siltstone (Larese, 1974).

The siltstone-sandstone blanket of western West Virginia is generally thinner than 12 m (40 ft), and the sandstone fines westward, merging with the thick Berea Siltstone of westernmost West Virginia, eastern Kentucky, and south-central Ohio (Fig. 20). Adjacent to the trends, the sandstone-siltstone blanket typically fines upward and contains quartz pebbles (Rittenhouse, 1946; Pepper et al., 1954). West of the trends, the sandstone generally coarsens upward and contains layers with quartz pebbles and some concentrations of heavy minerals (Larese, 1974; Potter et al., 1983). In core, Larese (1974) observed a bryozoan, and Potter et al. (1983) observed wave ripples.

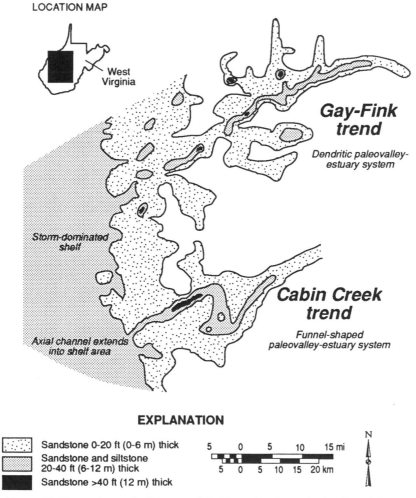

Figure 19. Net sandstone isolith map of the Berea Sandstone in the Gay-Fink and Cabin Creek trends, West Virginia (after Larese, 1974). The Gay-Fink and Cabin Creek trends are interpreted to represent paleovalley-estuary systems.

Environmental interpretation. Pepper et al. (1954) recognized the mature, quartzose composition of all sandstone in the Bedford-Berea sequence, including pebbly sandstone. Plotting sandstone composition on the QFL and Qm-F-Lt diagrams of Dickinson et al. (1983) suggests that the sandstone was derived from recycled orogenic sources (Fig. 21). However, the total lithic fraction is dominated by quartzose rock fragments such as polycrystalline quartz and chert, and the best evidence for ultimate derivation from an orogenic terrane is rock fragments with a schistose texture. Indeed, the mature quartzose nature of the pebbly sandstone indicates provenance primarily by recycling older sediment and sedimentary rock. The mineralogic similarity of Cussewago and Berea sandstone suggests a closer genetic link than had previously been supposed, and both appear to have been recycled mainly from the Catskill clastic wedge.

The pebbly-sandstone lithofacies is interpreted to represent the major fluvial systems of the Bedford-Berea sequence that eroded sediment of the Catskill clastic wedge and delivered sediment to the western basin. In addition to fluvial deposits, however, the lithofacies includes a deltaic system in the Cussewago–Second Berea Sandstone, and shelf, beach, and estuary deposits associated with the Gay-Fink and Cabin Creek trends. The following discussion identifies the nature and location of the principal Bedford-Berea sand sources and suggests that marine transgression and estuary formation were major factors in the alluviation and preservation of the coarse-grained sediment axes.

Cussewago–Second Berea Sandstone. Interpretation of the pebbly-sandstone lithofacies on the basis of outcrop data is limited because of poor exposure. Coarse grain size and crossbedding indicate vigorous sedimentation in a system rich with bedload, and the few available directional measurements suggest unidirectional flow toward the northwest (Fig. 16). The localized nature of the pebbly sandstone suggests deposition in fluvial channels that scoured the Riceville and Chagrin Formations, but the best clues to the origin of the pebbly-sandstone lithofacies are in the subsurface.

In the subsurface the Cussewago–Second Berea Sandstone contains several distinctive paleogeographic elements. The linear geometry of the northwest-trending sandstone axis in Lawrence

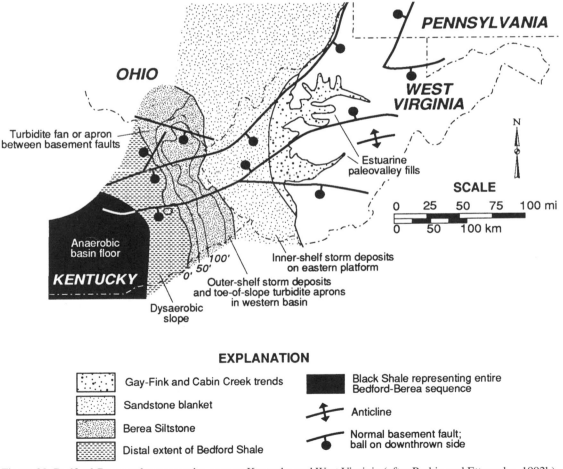

EXPLANATION

Gay-Fink and Cabin Creek trends		Black Shale representing entire Bedford-Berea sequence	
Sandstone blanket		Anticline	
Berea Siltstone		Normal basement fault; ball on downthrown side	
Distal extent of Bedford Shale			

Figure 20. Bedford-Berea paleogeography, eastern Kentucky and West Virginia (after Pashin and Ettensohn, 1992b). Sand-rich estuary and shelf deposits predominated on the eastern platform, and silt- and mud-rich shelf, slope, and basinal environments predominated in the oxygen-deficient western basin. Structural control of sedimentation is apparent by preservation of the branching paleovalley-estuary deposit of the Gay-Fink trend between major basement faults in West Virginia and by local deflection of isopach contours in the Berea Siltstone of eastern Kentucky.

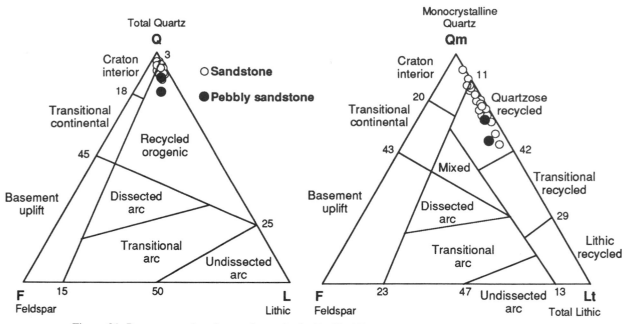

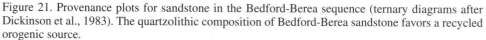

Figure 21. Provenance plots for sandstone in the Bedford-Berea sequence (ternary diagrams after Dickinson et al., 1983). The quartzolithic composition of Bedford-Berea sandstone favors a recycled orogenic source.

County, Pennsylvania (Figs. 12, 17), verifies the fluvial hypothesis of Pepper et al. (1954). The northwest orientation of the main Cussewago sediment axis suggests sediment sources in eastern Pennsylvania, Maryland, and northern Virginia. Schumm (1977) asserted that only "big rivers," such as the Nile and the Mississippi, have channels deeper than 20 m (65 ft), and indicated that thicker sandy successions typically represent alluvial valley fills. Presence of the thickest Cussewago–Second Berea Sandstone between the Transylvania fault and the Tyrone–Mount Union lineament suggests loose control of the position of the main alluvial valley by basement structure (Fig. 22).

Bifurcation of the main sandstone axis and the lobate geometry defined by the 50-ft (15-m) contour in northwestern Pennsylvania (Fig. 12) indicate deltaic environments—this is the Cussewago delta of Pepper et al. (1954). Deltaic successions are classically characterized as coarsening upward (e.g., Coleman, 1982), but the Cussewago–Second Berea is a fining-upward sequence that generally rests erosionally on the Chagrin and Riceville Formations (Fig. 18). In the Bedford-Berea sequence, this relationship reflects development of the subaerial delta on the eastern platform and accommodation of distal deltaic and open-marine sediment by the western basin.

The sharp, convex base of the Cussewago–Second Berea Sandstone (Figs. 7–9), which causes general thickening of the Bedford-Berea sequence in northeastern Ohio (Fig. 11), suggests that the sandstone eroded underlying Catskill strata. At the scale of individual bifurcating sandstone trends, however, thickening of the Cussewago–Second Berea coincides only

locally with thickening of the Bedford-Berea. Therefore, the bifurcating trends are interpreted to represent more than one cycle of establishment, abandonment, and reworking of distributaries. Some abandoned and reworked distributary systems are best viewed on the Bedford-Berea isopach map (Fig. 11) and thus probably incised the Chagrin Shale; they include the isopach maximum in Trumbull County, which is directed toward the northeast part of the western basin, and the minor axis in Columbiana County (Fig. 12).

Linear thickening of the Bedford-Berea sequence in the southern part of the hooklike form (Figs. 11, 12) is interpreted to represent early development of a major channel system, and the sharp base and general fining upward of the Cussewago–Second Berea (Figs. 8–10, 18) support this interpretation. However, parts of the Cussewago–Second Berea appear to be built above rather than scoured into underlying sediment (Fig. 9), suggesting later formation of delta-destructive barrier-island arcs as is now occurring on the Mississippi Delta (Penland and Boyd, 1985). Identifying where the subaerial delta gave way to marine environments is difficult, but sandstone distribution indicates that the delta plain was mainly north of the faults in northeastern Ohio and may have had an elongate southward extension that fed the Cussewago–Second Berea siltstone belt (Figs. 12, 17).

Gay-Fink and Cabin Creek trends. The branching geometry (Fig. 19) and the fining-upward sequence of the Gay-Fink trend are typical of contributive fluvial systems, and the presence of the transgressive, basinal Sunbury Shale at the top of

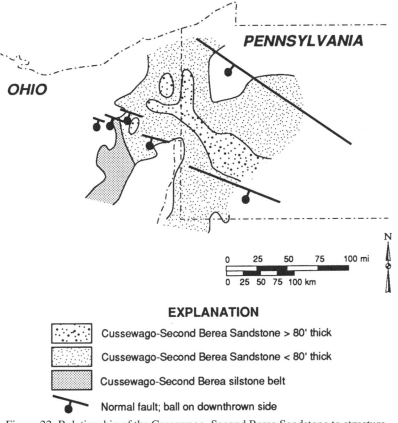

Figure 22. Relationship of the Cussewago–Second Berea Sandstone to structure, northeastern Ohio and northwestern Pennsylvania.

EXPLANATION

Cussewago-Second Berea Sandstone > 80' thick

Cussewago-Second Berea Sandstone < 80' thick

Cussewago-Second Berea silstone belt

Normal fault; ball on downthrown side

the sequence suggests that the trend is partly an estuary deposit. Branching estuaries, such as Chesapeake Bay, form by transgression and aggradation in deeply incised paleovalleys (Fairbridge, 1980). The Cabin Creek trend, by contrast, is interpreted to represent a funnel-shaped estuary. Funnel-shaped systems, such as the Gironde estuary of France, typically form in response to inundation of transitive coastal plain channels and have a lower gradient than branching systems (Fairbridge, 1980).

Because estuaries form in drowned river valleys, the Gay-Fink and Cabin Creek fluvial systems probably extended farther west than is now apparent. Indeed, the axial sandstone of the Cabin Creek trend extends southwestward into the siltstone-sandstone blanket (Fig. 19), suggesting that the systems had a constructive phase prior to estuary development. Larese (1974) suggested that beach deposits are at the west end of the Gay-Fink trend on the basis of heavy-mineral and quartz-pebble concentrations. He suggested that beaches formed as channel sand was reworked during transgression.

Bounding of the Gay-Fink trend by the faults of the Rome trough suggests strong structural control of sedimentation (Fig. 20), as has already been suggested by Larese (1974) and Boswell (1988). The branching, contributive nature of the Gay-Fink system indicates that considerable topographic relief was associated with the faults; the tributaries probably had local

sources in upthrown fault blocks, whereas the trunk channel probably had a source east of the cross-over rift. The funnel shape of the Cabin Creek trend indicates limited topographic relief. The eastern limit of the trunk channel is near the Pocono dome, suggesting that the trend drained sediment eroded from the structure.

Quarrystone lithofacies

Characteristics. Sandstone of the quarrystone lithofacies is restricted to the Berea of north-central Ohio and is best known from the type area in Berea, Cuyahoga County, and from the sandstone quarries of the region, which quarrymen boast to be the deepest in the world (Fig. 23). The quarrystone and the closely related deformed silty-sandstone lithofacies are not separable on regional maps and cross sections and are therefore combined in Figure 7. The quarrystone is discontinuous and, although the full thickness of the quarrystone is not exposed, the sandstone is a maximum of 75 m (250 ft) thick in the famous Buckeye quarry at the Cleveland Quarries Company (locality 23).

Thickness of the quarrystone is extremely variable throughout north-central Ohio, and the sandstone becomes increasingly discontinuous toward the west. Where exposed,

Figure 23. Photograph of the Harrison Road quarry (locality 20). Berea sandstone quarries in north-central Ohio are among the deepest in the world. Berea quarrystone is used primarily for building stone and rip rap and is also considered a standard for reservoir engineering tests by petroleum companies (Potter et al., 1983). The Harrison Road quarry contains approximately 24 m (78 ft) of sandstone containing dominantly horizontal laminae with parting lineation, which is the most desirable building stone.

ing topographic escarpment in the eastern part of the Oberlin Quadrangle, Lorain County; quarries of the Cleveland Quarries Company (locality 23) were excavated along the escarpment. West of the escarpment, quarries were excavated into elongate topographic highs that may be used to map the quarrystone.

The quarrystone is thickly bedded and contains few shale beds or partings. The sandstone contains exclusively unidirectional bedforms, including crossbeds, current ripples, and horizontal laminae; only one or two types of bedform are present at any given exposure. The crossbeds are generally tangential (Fig. 24B), and grouped trough and planar forms abound. Individual bedsets are as thick as 3 m (10 ft), and in some exposures, solitary crossbeds are present. Climbing ripples, or ripple-drift structures, are the most common type of current ripple, and lunate and linguoid types are prevalent (Fig. 24C,D). Horizontally laminated sandstone, which is the most desirable building stone, has well-developed parting lineation.

Conglomeratic beds containing well-rounded pebbles and cobbles of red sideritic siltstone with limonitic rinds and gray and black shale clasts are exposed at a few localities and are most abundant near the lower contact of the sandstone. Platy vitrain clasts as large as cobbles are locally abundant at the base of the sandstone and are present sparingly throughout. Conglomeratic zones containing well-rounded quartz pebbles, bluish-black chert pebbles, and metamorphic "greenstone" pebbles are exposed at the Deep Lock quarry (locality 38). Unabraded shark teeth and a bryozoan were also observed at the Deep Lock quarry (Fig. 24E).

Because strata are generally exposed in vertical rock faces, few directional readings are available from the quarrystone lithofacies. Even so, the available paleocurrent data are unidirectional, and the vector magnitude at localities where more than one reading was taken exceeds 0.90. At individual localities, moreover, readings taken on different types of sedimentary structures are similar. Most of the variation in the vector azimuths is regional (Fig. 16). In the western part of the quarrystone lithofacies, the vector azimuths are directed west and range from 248° to 260°. Farther east, the vector azimuths are directed northwest and range from 285° to 311°. These results agree with the more detailed paleocurrent study of Lewis (1988).

The quarrystone contains large-scale deformational structures bounded by large, contorted shale masses (Coogan et al., 1981; Lewis, 1988). Microfaults are abundant near the corners and sides of the sandstone quarries, and Pepper et al. (1954) demonstrated that the sandstone is absent only meters away from the quarry margins. Individual faults penetrate no more than 2–3 m (6–10 ft) of sandstone and are truncated by all manner of bedforms and scour-and-fill structures.

Sandstone layers in most quarries are slightly tilted. The steepest tilt is defined by a shaly bed in quarry 7 at the Cleveland Quarries Company (locality 23), where the beds dip approximately 10° (Fig. 24F). The layers in each quarry have a different degree and direction of dip, and some dips are

the lower contact of the quarrystone lithofacies is sharp and irregular, and the quarrystone locally truncates subjacent strata (Fig. 24A). Much of the irregularity owes to normal faulting, and most faults have net displacement less than 2 m (6 ft) (Fig. 24B). The upper contact is exposed only at Berea (locality 28), where strata of the solitary-crossbed lithofacies separate the quarrystone from the Sunbury Shale. Quarrystone strata also are present in Summit County, far removed from other exposures of the lithofacies (Fig. 7).

Geomorphology provides key evidence for sandstone-body geometry. The quarrystone blanket breaks up into a series of isolated, elongate bodies that are readily apparent in topographic quadrangle maps (Fig. 25). Passage of the blanket into the elongate bodies is defined by a subtle, north–south-trend-

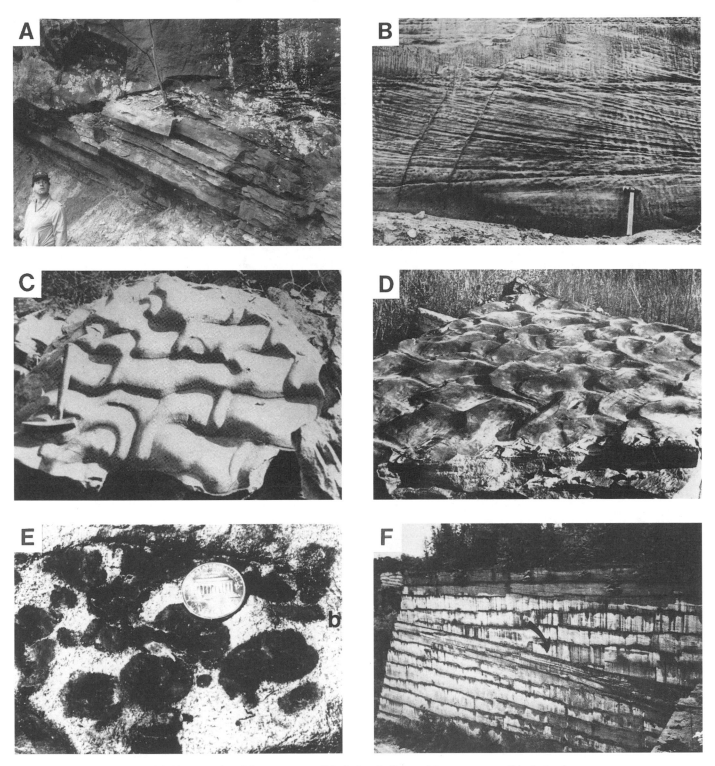

Figure 24. Photographs of the quarrystone lithofacies. A, Strata of the quarrystone lithofacies truncate dipping beds of the deformed silty-sandstone lithofacies at Elyria (locality 25). B, Tangential cross-bed with microfaults at the Cleveland Quarries Company (locality 23) (after Potter et al., 1983). Faults become increasingly common toward the margins of the quarry sandstone bodies. C, Linguoid ripples at the Cleveland Quarries Company (after Pepper et al., 1954). D, Lunate ripples at the Cleveland Quarries Company (after Potter et al., 1983). E, Ironstone conglomerate containing bryozoan fossil (b) at base of quarrystone lithofacies, Deep Lock Quarry Park (locality 38). F, Dipping shale bed (arrow) in quarry face, Cleveland Quarries Company (after Potter et al., 1983). Quarry face is 26 m (85 ft) high.

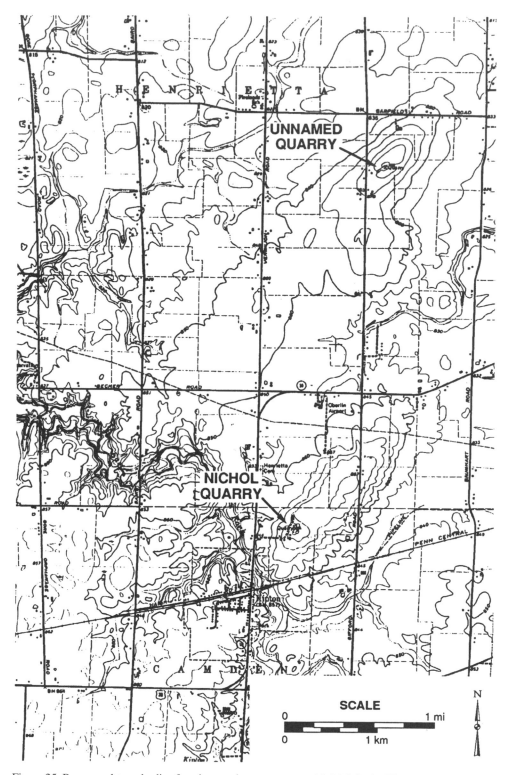

Figure 25. Berea sandstone bodies forming an elongate topographic high in the Kipton 7.5-min Quad-rangle, Lorain County, Ohio. The quarrystone has been mined near the northeast end of this landform, whereas strata of the deformed silty-sandstone lithofacies are exposed near the southwest end.

opposed. However, precise determination of dip direction is difficult because the sandstone is exposed in inaccessible, vertical quarry faces. At Elyria (locality 25), bedding within the quarrystone is more or less horizontal, but the sandstone is intruded by a diapiric mass of black Cleveland Shale. On the north side of the diapiric mass, the sandstone is dominated by ripple-drift cross-laminae with some solitary crossbeds, whereas on the south side, the sandstone is dominated by grouped trough crossbeds with minor lenses of ripple-drift cross-laminae. Downstream from the quarrystone at the Rocky River Reservation (locality 27), the gray phase of the Bedford (gray shale lithofacies) is deformed into a small, overturned fold that overlies a series of thrust faults in Bedford and Cleveland shale (Fig. 26).

The detailed isopach map of Pepper et al. (1954) indicates that sandstone thickness changes in the area of the Middleburg fault (Fig. 27). West of the fault in the downthrown block the thickness of the Berea varies extremely, as is characteristic of the quarrystone. East of the fault, the Berea is more uniform in thickness and contains quarrystone strata in only a few locaions; exposures east of the fault are mainly of the solitary crossbed lithofacies, which is discussed in a later section.

A core from Ashland County, north-central Ohio, where Pepper et al. (1954) identified extreme thickness variation in the Berea, contains convoluted silty sandstone with slickensided shale partings, which grades upward into sandstone (Fig. 18). The upper portion of the core contains of 52 m (172 ft) of yellowish-brown silty sandstone with abundant current-ripple cross-laminae, horizontal laminae, and gently inclined laminae. Although the rock is fine grained and lacks large-scale crossbeds, dominance of current-ripple cross-laminae suggests that it belongs to the quarrystone lithofacies. The general distribution of the lithofacies is in an area variable sandstone thickness bounded by the 50-ft (15-m) contour in the western basin of Ashland, Richland, Medina, Huron, and Lorain Counties (Figs. 17, 28).

Environmental interpretation. Although the quarrystone lacks distinct modern analogs, abundant sedimentary and soft-sediment deformation structures make it one of the most dynamic lithofacies to interpret. The discussion that follows supports the idea put forth by Coogan et al. (1981) and Lewis (1988) that subsidence into soft mud caused by loading of sand enabled rapid accumulation of unusually thick sandstone sequences. The quarrystone was evidently derived from the east and southeast and is interpreted to have accumulated in the proximal part of the Cussewago delta front.

The quarrystone is unique because as much as 76 m (250 ft) of sandstone is present (Fig. 23) and, in all but one exposure, the full thickness of the sandstone lacks significant shale beds. The numerous normal faults within the sandstone clearly are synsedimentary, because they are truncated by bedforms and scour-and-fill structures. The geometry of the quarrystone-Bedford contact indicates that the quarrystone bodies are essentially fault-bound blocks (Fig. 29). Therefore, rapid subsidence along faults caused by loading of sand on soft mud may have enabled such a great thickness of sand to accumulate. Uneven subsidence and downslope creep of the fault-bounded sandstone blocks may have contributed to the variable bedding tilt

Figure 26. Bedford Shale (gray shale lithofacies) is intensely deformed and contains folds and thrust faults (t) adjacent to quarry sandstone bodies at the Rocky River Reservation (locality 27). A, Shale and siltstone with shear fractures. B, Intensely sheared gray shale. C, Overturned fold in shale and siltstone. Field of view is approximately 5 m (16 ft) tall. These deformation structures apparently formed as subsiding quarrystone bodies pushed sediment aside.

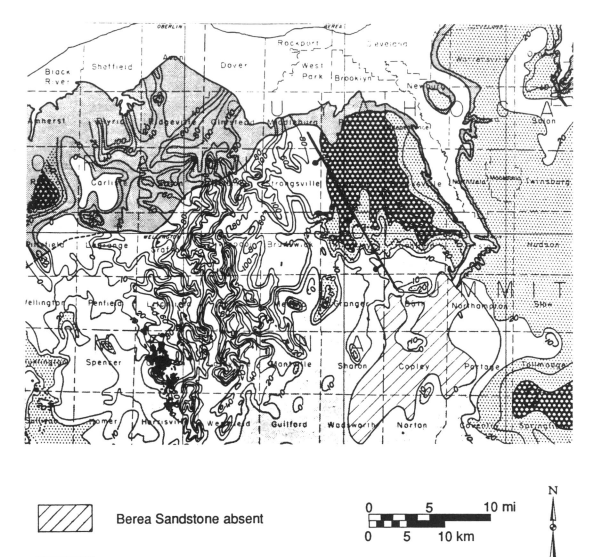

Berea Sandstone absent

Berea Sandstone < 20 ft (6 m) thick

Berea Sandstone 20-60 ft (6-18 m) thick

Berea Sandstone > 60 ft (18 m) thick

Middleburg fault

Figure 27. Isopach map of the Berea Sandstone in the area of the Middleburg fault (adapted from Pepper et al., 1954). The quarrystone lithofacies, which is characterized by extreme thickness variation, makes up the Berea Sandstone southwest of the Middleburg fault, whereas the solitary-crossbed lithofacies, which is characterized by modest thickness variation, makes up most of the Berea northeast of the fault.

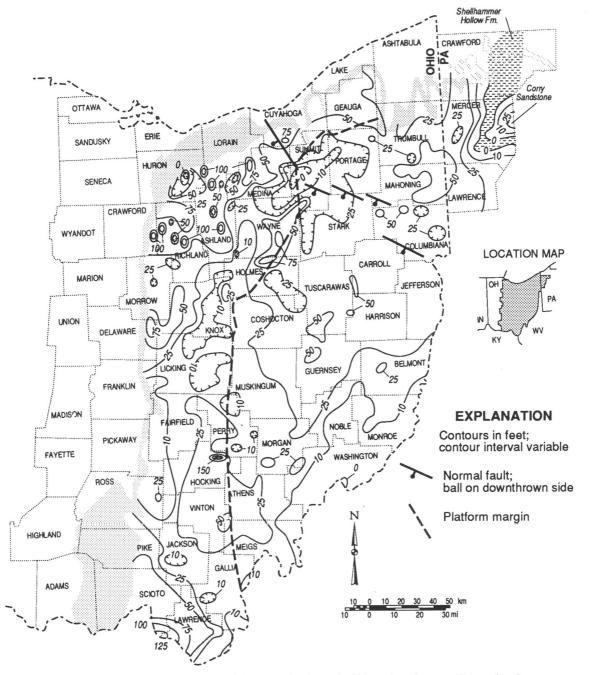

Figure 28. Net sandstone isolith map of the Berea Sandstone in Ohio and northwestern Pennsylvania.

seen in most quarries (Fig. 24F) and apparently pushed sediment forward as well as laterally, thereby forming thrust faults and overturned folds (Fig. 26).

Bedforms (Fig. 24) and paleocurrent data (Fig. 16) from the quarrystone indicate strong unidirectional flow, and the normal absence of shale partings suggests that the flows were more or less permanent. The general lack of lateral change in the bedforms within the sandstone bodies indicates that sand was deposited primarily by sheet flow. However, scarce scour-and-fill structures indicate that flow was locally channelized.

Paleocurrent evidence from the eastern outcrops of the quarrystone indicates that a considerable amount of sand was transported northwest, parallel to the Middleburg fault. The detailed isopach map of Pepper et al. (1954) further indicates control of sandstone thickness by the fault (Fig. 27). Wave-ripple orientation in the deformed silty-sandstone lithofacies (Fig. 16), which is intimately related to the quarrystone and is discussed in the next section, also indicates a west-dipping paleoslope. In the western reaches of the quarrystone, however, paleocurrent indicators indicate flow toward the west and south-

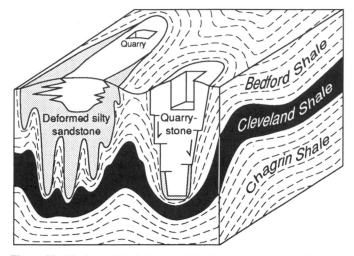

Figure 29. Block model of elongate delta-front sandstone bodies containing the quarrystone and deformed silty-sandstone lithofacies.

west, suggesting capture of currents by the prevailing regional paleoslope.

Coalified plant debris in the quarrystone suggests that unidirectional, sediment-laden flows originated in terrestrial environments, although bryozoan fossils (Fig. 24E) and shark teeth indicate that the flows acted in marine environments. West and northwest paleocurrent data (Fig. 16) suggest that the quarrystone was not derived mainly from the north, as suggested by Pepper et al. (1954), but was derived mainly from the east or southeast, as suggested by Lewis (1968, 1976, 1988) and Coogan et al. (1981).

Paleocurrent data suggest that currents may have originated in the Cussewago delta. Additionally, the Mahoning County distributary terminates less than 15 mi (24 km) from the quarrystone outcrops in Summit County (localities 37, 38) (Fig. 17). Hence, the quarrystone is interpreted to be the proximal part of the Cussewago delta front. Another connection between the quarrystone and the Cussewago delta may have existed at the northeast end of the western basin but, as discussed below, that connection was evidently reworked late in Berea deposition.

Deformed silty-sandstone lithofacies

Characteristics. In north-central Ohio, silty sandstone contains the most complex and diverse deformation structures in the Bedford-Berea sequence; these structures constitute the deformed silty-sandstone lithofacies and include grand-scale ball-and-pillow structures, complex folds, normal and thrust faults, and overturned bedding. The lithofacies crops out amidst the Berea Sandstone quarries of Erie, Huron, Lorain, and western Cuyahoga Counties (Fig. 7) but is not economic in this area. Where present, the deformed silty-sandstone lithofacies forms the base of the Berea.

The deformed silty sandstone contains abundant wave ripples (Fig. 30A), and current-ripple-drift cross-laminae are present locally. Where more or less horizontal, the silty sandstone is in fairly even, thin to thick beds separated by laminae and thin beds of shale. Sparse feeding burrows are in the sandstone. Wave-ripple azimuths have high vector magnitudes ranging from 0.97 to 1.00. In the western part of the lithofacies, the ripple crests strike north with vector azimuths ranging from 335° to 0° (Fig. 16). Farther east, however, the azimuths turn northwest, and at the easternmost locality (Olmsted Falls, locality 26), the wave ripples have a vector-mean azimuth of 306°, which is typical of other parts of the Bedford-Berea sequence.

In the eastern outcrops of the deformed silty sandstone, the lithofacies rests on the red phase of the Bedford Shale (red shale lithofacies) and ranges from 1–5 m (3–16 ft) in thickness. Strata become discontinuous farther west and are truncated by the quarrystone at Elyria (locality 25) (Fig. 24A). Deformation structures in the Elyria area include ball-and-pillow structures, load casts, microfaults, and open folds. These structures increase westwardly in size and abundance and, at Olmsted Falls (locality 26), ball-and-pillow structures are locally more than 4 m (13 ft) thick and more than 8 m (26 ft) wide (Fig. 30B).

West of Lorain (locality 24), the deformed silty-sandstone lithofacies forms localized, intensely deformed masses thicker than 10 m (30 ft) that overlie the gray phase of the Bedford (gray shale lithofacies) (Fig. 31). The upper contact of the deformed silty sandstone is scarcely exposed in this area, so additional tens of meters may be present. The lower surfaces of the bodies are convex-side downward, and the bodies are bounded on the sides by faults or nearly vertical beds. Each sandstone body has unique characteristics ranging from clusters of convoluted and microfaulted silty sandstone balls to overturned sheet sandstone beds with wave ripples (Fig. 30C).

Adjacent to the bodies, gray shale has cleavage that conforms to the margins of the silty sandstone masses, but decimeters away from the bodies, the shale has poorly oriented cleavage and commonly contains convoluted, cobble- to boulder-size silty sandstone balls (Fig. 30D). Red shale is generally exposed in deformed masses near the silty sandstone bodies (Fig. 31) but is not in contact with the bodies. Near the silty sandstone masses, however, red shale is locally in fault contact with gray shale.

Stream transects reveal the large-scale aspects of the deformation structures in the westernmost outcrops and establish that the deformed silty sandstone masses are spaced variably both locally and among localities (Fig. 31). Where silty sandstone masses are widely spaced, gray Bedford shale thins sharply beneath the deformed bodies. The black Cleveland Shale is deformed into tight synclines below the sandstone bodies and, between bodies, the Cleveland is folded into broad anticlines, and the Bedford-Berea contains little internal deformation. Along the Vermillion River (locality 22) some of the anticlines are asymmetric, and the Olmsted Member of the Ohio Shale has been exhumed only a short distance from the deformed sandstone masses.

The relationship between the deformed silty sandstone and the quarrystone lithofacies is unclear in the western outcrops

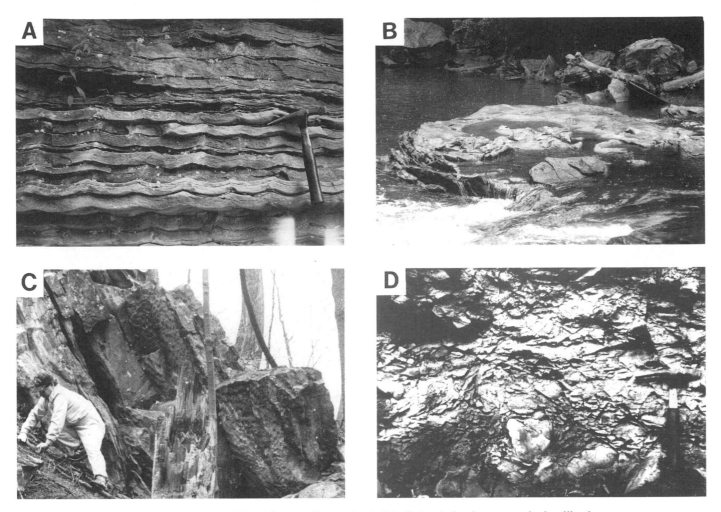

Figure 30. Photographs of the deformed silty-sandstone lithofacies. A, In-phase wave ripples, like those at Old Woman Creek (locality 18), are characteristic of the deformed silty-sandstone lithofacies. B, Ball-and-pillow structures more than 8 m (26 ft) wide at Olmsted Falls (locality 26). C, Overturned beds with interference wave ripples at Chappel Creek (locality 19). D, Bedford shale (gray shale lithofacies) with poorly oriented cleavage and convoluted sandstone balls near margin of deformed silty sandstone mass.

because a topographic discrepancy of more than 45 m (150 ft) exists between the stream exposures and the quarries; thus, a significant portion of the two lithofacies is not exposed. At Old Woman Creek (locality 18), however, more than 10 m (33 ft) of deformed silty sandstone crops out at the southwest extremity of one elongate topographic high, and a sandstone quarry is at the northeast extremity. This configuration suggests that, in some elongate bodies, the quarrystone passes southwestward into deformed silty sandstone (Fig. 29). The two cores from the Berea of Ashland County contain current ripples, wave ripples, burrows, and microfaults, and thus have characteristics of both the quarrystone and deformed silty-sandstone lithofacies (Fig. 18). Perhaps these cores are characteristic of strata transitional between those in the stream exposures and the quarries.

Subsurface cross sections demonstrate that black Cleveland Shale is downwarped below thick sandstone bodies in north-central and central Ohio (Figs. 8–10). Pepper et al. (1954)

mapped a series of thick sandstone bodies that are aligned along a north-south trend between Ashland and Hocking Counties, but these bodies are thinner than the others and were not penetrated by wells with gamma ray logs. Nevertheless, subsurface data indicate that sandstone bodies of the quarrystone and deformed silty-sandstone lithofacies extend south along the platform margin from Ashland County to Hocking County.

Environmental interpretation. Whereas the quarrystone is interpreted to represent the proximal Cussewago delta front in which unidirectional flow, faulting, and subsidence of sand into soft mud were the dominant depositional processes, the deformed silty sandstone is interpreted to represent the distal Cussewago delta front in which wave action, liquefaction, and diapirism were the dominant processes (Fig. 29). The erosional contact between the quarrystone and the deformed silty sandstone (Fig. 24A) is typical of delta-front environments because distributary and proximal mouth-bar deposits commonly trun-

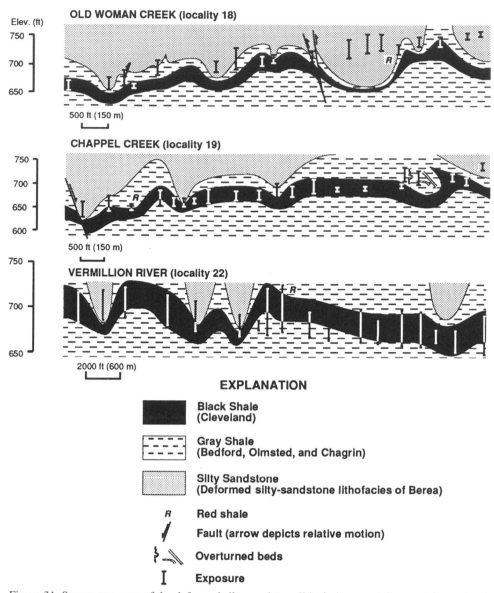

Figure 31. Stream transects of the deformed silty-sandstone lithofacies reveal diverse deformational styles.

cate distal-bar deposits (Fisk, 1961). However, the preponderance of wave ripples in the silty sandstone (Fig. 30A) is more characteristic of shoal-water mouth bars than of classical river-dominated mouth-bar deposits (Elliott, 1975). Therefore, the Cussewago delta may best be interpreted as an extensive shoal-water deltaic system.

Diapirism and associated mass movements have been recognized as significant factors that affected deposition of the deformed silty sandstone and quarrystone (Wells et al., 1991). Diapiric structures are common in delta-front sequences where sand overlies thick, undercompacted shale successions (O'Brien, 1968; Morgan et al., 1968; Bruce, 1973), and large-scale ball-and-pillow structures (>3 m in diameter) (Fig. 30B) also are common in delta-front sequences (Howard and

Lohrengel, 1969). Superposition of sand on mud gives rise to an unstable, inverse density gradient that may be stabilized by intrusion of mud into the overlying sand, as well as by foundering of sand into the mud (O'Brien, 1968). In the Mississippi Delta, mud diapirism is an ongoing process, and deformation of mud masses includes thrust faults and overturned bedding (Morgan, 1961; Morgan et al., 1968).

Overturned bedding (Fig. 31) suggests that deformation structures in the deformed silty-sandstone lithofacies are as complex as the Mississippi Delta diapirs. Like the quarrystone, the great thickness of the deformed silty sandstone may be explained by subsidence of sand into soft mud. However, deformation of the silty sandstone was different from that of the quarrystone, because convolute bedding indicates that a major component of liquefaction was associated with the deformation. For example, many of the large, deformed silty sandstone masses resemble ball-and-pillow structures more closely than fault-bounded blocks. Although the deformed shale masses in the Bedford resemble mud diapirs, identifying the shale masses as true piercement structures is equivocal because only the broad, lower parts of the masses are exposed.

Deformation structures resembling those in the deformed silty-sandstone lithofacies apparently are widespread in the northern part of the western basin, because downwarped black shale was detected below the thickest sandstone bodies in all gamma ray logs (Figs. 8–10). This configuration indicates that the sandstone bodies were not deposited in simple channels as proposed by Pepper et al. (1954), but that they formed primarily by subsidence of sand into soft mud. Furthermore, the thick, localized body of silty sandstone in northern Hocking County demonstrates that subsiding sand bodies extended south along the eastern margin of the western basin into central Ohio.

Elongate sandstone bodies containing the quarrystone and deformed silty sandstone superficially resemble fan-shaped to elongate delta-front sandstone bodies such as those described by Fisk (1961). However, obliquity of paleocurrent vectors to the sandstone-body axes indicates that the elongate bodies did not form as small fans or channelform conduits. Moreover, consistency of paleocurrent indicators within and among the elongate bodies indicates a much more uniform paleoslope than

would be expected in a delta front containing fans and channels. Therefore, the elongate geometry is interpreted to have formed mainly by faulting, diapirism, and subsidence of sand into soft mud, and may have had little relationship to surface topography at the time of deposition. Along these lines, Coogan et al. (1981) suggested that some deformed sandstone bodies have more affinity with large-scale bars than with channels.

Low-angle crossbed lithofacies

Characteristics. In central Ohio, the Berea Sandstone contains sandstone and silty sandstone assigned to the low-angle crossbed lithofacies (Fig. 32). Sandstone is restricted to the northern part of the lithofacies. The lower contact is exposed only at Mount Gilead (locality 14), where the sandstone sharply overlies the Bedford Formation with a planar contact. The upper contact of the sandstone is not exposed, and thus facies thickness is unknown; a maximum thickness of about 3.25 m (11 ft) is exposed at the Galion Reservoir (locality 16) and at the Leesville quarry (locality 17). The most common sedimentary structures are low-angle, wedge-planar crossbedding and horizontal laminae; both are in thick beds having a platy to flaggy parting. At the Galion Reservoir (locality 16), the sandstone contains hummocky cross-strata, and wave ripples were found on one bedding plane near the top of the section. Sandstone in the low-angle crossbed lithofacies lacks fossils and bioturbation.

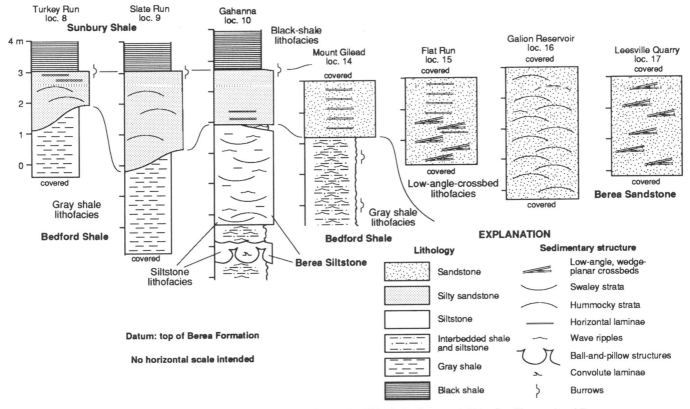

Figure 32. Measured sections of the low-angle crossbed lithofacies in central Ohio. See Figures 4 and 7 for locations; no horizontal scale intended.

At the top of the Berea, Hyde (1953) identified a thick, silty sandstone bed ranging in thickness from 1–3 m (3–10 ft) that is widespread in central Ohio and makes up the southern part of the lithofacies (Fig. 32; localities 8–10). This silty sandstone bed is recognized only in the central part of the state, but Hyde suggested that it could be correlated with a thick siltstone bed that forms the top of the Berea in southern Ohio. At Turkey Run and Slate Run (localities 8, 9), this bed forms the entire Berea and rests upon the gray shale lithofacies of the Bedford. Here, the base of the stratum is sharp and broadly undulatory, accounting for most of the thickness variation. Farther north at Gahanna (locality 10), the bed is planar and overlies several meters of Berea Siltstone. The bed generally appears massive, but faint hummocky strata and horizontal laminae are locally visible. The uppermost part of the layer is bioturbated and contains marcasite nodules; the bed is overlain sharply by black Sunbury Shale.

The subsurface distribution of the low-angle crossbed lithofacies is questionable. The sandstone bed making up the northern part of the lithofacies can be traced into the area of variable sandstone thickness containing the quarrystone and deformed silty-sandstone lithofacies (Figs. 17, 33). Sandstone of the low-angle crossbed lithofacies is probably dispersed among the deformed bodies, and shale and siltstone in the Bedford Formation (gray shale lithofacies) is folded and faulted at Mount Gilead (locality 14).

The silty sandstone bed extends into central Ohio where the Berea is approximately 3 m (10 ft) thick in the nearby subsurface (Fig. 28). A widespread blanket of siltstone thinner than 8 m (25 ft) and locally thinner than 3 m (10 ft) extends from Wayne County, north-central Ohio, to Lawrence County, south-central Ohio, and to Washington County, southeastern Ohio, and may include the low-angle crossbed lithofacies (Fig. 33); that blanket forms the top of the Bedford-Berea sequence (Figs. 9, 10).

Environmental interpretation. Since the classic study of Thompson (1937), most low-angle, wedge-shaped crossbedding in sandstone has been associated with beach environments, particularly the swash and surf zones (e.g., Clifton et al., 1971; Hunter et al., 1979). The hummocky cross-stratified sandstone at the Galion Reservoir (Fig. 32; locality 16) is typical of storm-wave deposits (Dott and Bourgeois, 1982; Swift et al., 1983). Although hummocky strata have also been identified in surf zone deposits, they are interspersed with low-angle, wedge-planar crossbeds (Greenwood and Sherman, 1986). Because wave ripples accompany the hummocky strata, this part of the low-angle crossbed lithofacies probably formed in a shoreface or inner shelf setting. Hummocky strata in silty sandstone of the low-angle crossbed lithofacies also suggests a storm-related origin, and the apparent regional extent of the layer (Figs. 32, 33) indicates deposition in shelf environments.

Sandstone in the low-angle crossbed lithofacies lacks the typical coarsening-upward sequence associated with prograding coastal barriers. Indeed, the lithofacies sharply overlies the

Bedford Shale at Mount Gilead (Fig. 32; locality 14), suggesting a disconformable lower contact. Apparent dispersal of the sandstone among the deformed sandstone and shale bodies of north-central Ohio suggests that possible mudlumps and differential compaction of mud around deformed sand masses provided cores for beach accretion. Therefore, sandstone of the low-angle crossbed lithofacies is interpreted to have been derived from destructive reworking of the Cussewago delta front by waves and storms late in Bedford-Berea deposition. A similar origin was postulated for delta-destructive sandstone bodies in the subsurface of Medina County (Burrows, 1988).

Solitary-crossbed lithofacies

Characteristics. The solitary-crossbed lithofacies contains distinctive sedimentary and deformational structures that contrast strongly with those in other Bedford-Berea strata. The lithofacies is the Berea Sandstone of eastern Cuyahoga, Geauga, Lake, Trumbull, and Ashtabula Counties, Ohio. Some of the facies was mapped as Cussewago Sandstone by de Witt (1951) and Shiner and Gallaher (1979) in northeastern Ohio and northwestern Pennsylvania, but the Berea Sandstone (i.e., solitary-crossbed lithofacies) extends above the Cussewago Sandstone into Pennsylvania (Fig. 34). The Berea Sandstone sharply overlies the Cussewago Sandstone (pebbly-sandstone lithofacies) and, farther east, the Berea overlies the Riceville Shale; west of the Cussewago, the Berea overlies a westwardly thickening wedge of the Bedford Shale. The solitary-crossbed lithofacies is overlain by black Sunbury Shale in Ohio and by gray shale and siltstone of the Bedford and Shellhammer Hollow Formations in Pennsylvania. Lithofacies thickness is 10–20 m (30–60 ft) in Ohio and is less than 5 m (16 ft) in northwestern Pennsylvania.

In much of northern Ohio, the basal few centimeters of the Berea Sandstone are cemented with marcasite. Above the marcasite, in the lower 2 m (6 ft) of the Berea, concretionary ferroan calcite and ferroan dolomite cement are common. In Pennsylvania, sandstone cemented with concretionary carbonate locally forms boulder-size spheroids, and marcasite is absent. Carbonate cement pervades some of the coarsest sandstone beds, which have been called siliceous limestone (White, 1880; Caster, 1934; Pepper et al., 1954).

The lithofacies is named for abundant high-angle (dip >15°), solitary crossbeds (Fig. 35). Grouped crossbeds are common, but cosets comprise only two or three bedsets (Fig. 34). Bedsets are typically between 0.5–3 m (1.5–10 ft) thick and are in places thicker than 7 m (23 ft) (Fig. 36). Planar-tabular and planar-tangential crossbeds predominate, and trough crossbeds are present locally. At Doan Brook, more than 3.7 m (12 ft) of low-angle, wedge-planar crossbedding is present near the base of the Berea (Fig. 37A). Bedsets tend to be uniform in thickness and generally extend for more than 100 m (330 ft); lensoid sets are present locally. Many crossbeds have reactivation surfaces, beyond which tabular crossbeds pass lat-

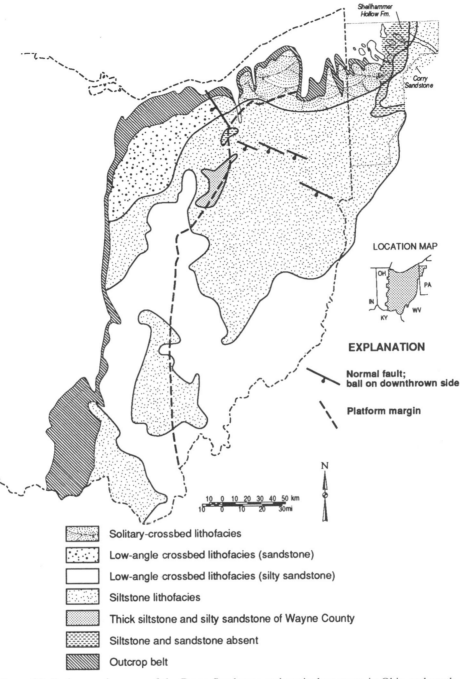

LOCATION MAP

EXPLANATION

——•—— **Normal fault;**
ball on downthrown side

— — — **Platform margin**

N

10 0 10 20 30 40 50 km
10 0 10 20 30mi

	Solitary-crossbed lithofacies
	Low-angle crossbed lithofacies (sandstone)
	Low-angle crossbed lithofacies (silty sandstone)
	Siltstone lithofacies
	Thick siltstone and silty sandstone of Wayne County
	Siltstone and sandstone absent
	Outcrop belt

Figure 33. Facies overlay map of the Berea Sandstone and equivalent strata in Ohio and northwestern Pennsylvania.

erally into tangential crossbeds (Fig. 38), some of which have clay drapes. One of the thickest crossbeds (Fig. 36) contains sigmoidal topsets. At Phelps Creek (locality 49) and the Bartholomew Quarry (locality 62), some crossbeds are truncated by asymmetrical, low-angle (dip <5°), trough crossbeds (Fig. 38) that are more than 6 m (20 ft) wide, are more than 0.3 m (1 ft) thick, and dip opposite the master bedsets.

Crossbed topsets are truncated by ripple-bedded sandstone with flaggy parting (Figs. 35, 36). Foresets commonly pass lat-

erally into rippled strata (Fig. 37B), and ripples are commonly superimposed on the toesets. Where rippled beds are a few meters thick, medium to thick lenses of tangential crossbedding are locally present; in places, rippled beds form low-angle, planar-tabular to planar-tangential crossbeds. Ripple wavelength generally ranges from only 1–4 cm (0.4–1.6 in), and ripple-crest orientation is extremely variable. The ripples are generally straight to sinuous crested (Fig. 37B), and flat-topped and interference wave and current forms are present (Fig. 37C).

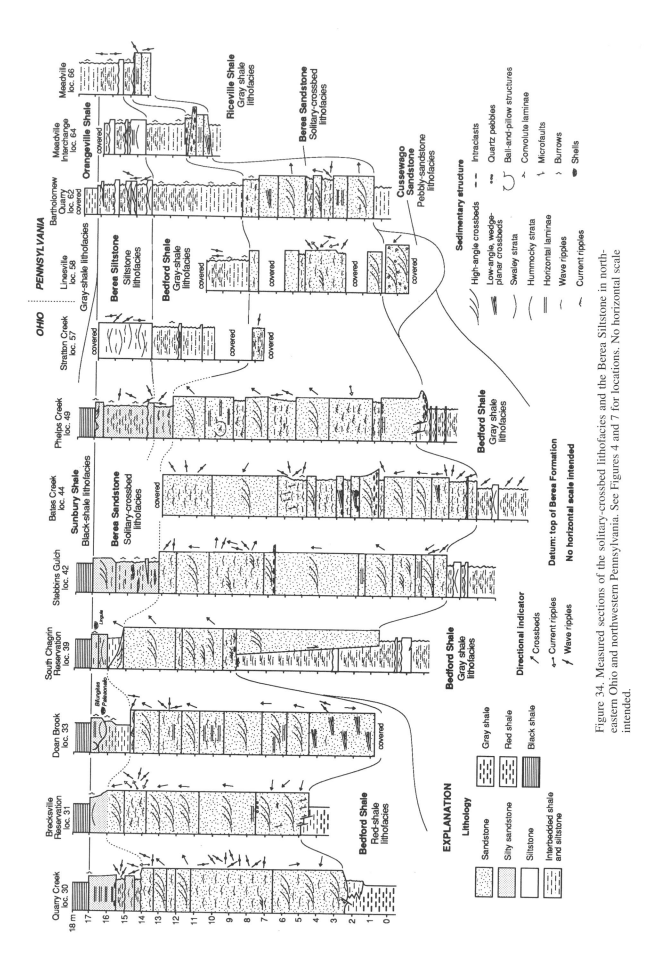

Figure 34. Measured sections of the solitary-crossbed lithofacies and the Berea Siltstone in northeastern Ohio and northwestern Pennsylvania. See Figures 4 and 7 for locations. No horizontal scale intended.

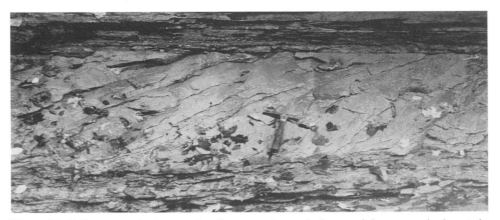

Figure 35. Solitary crossbed separating flaggy sandstone beds containing wave ripples at the Bartholomew Quarry (locality 62). Solitary crossbeds in the Berea are interpreted to represent migrating bars, and the intervening flaggy beds are interpreted to represent interbar sand-mud flats.

Figure 36. Crossbeds in the solitary-crossbed lithofacies reach a maximum thickness of 7 m (23 ft) at Stebbins Gulch (locality 42). This particular bedset has sigmoidal topsets suggestive of a tidal origin. Human scale in circle at bottom center of photograph.

Shale and sandstone with wavy, flaser, and lenticular bedding make up some flaggy intervals. Many shaly zones contain polygonal, sandstone-filled cracks (Fig. 37D), which are irregular in plan view and are in places clastic dike systems in which sand was injected from below (Fig. 37E).

Deformed sandstone is at the base of the solitary-crossbed lithofacies at many locations (Fig. 34). This sandstone is typically 3 m (10 ft) thick, rests sharply upon the Bedford Shale,

and is truncated with planar contact by undeformed solitary-crossbed strata. The basal contact typically is gently faulted or broadly folded, and the sandstone contains abundant normal microfaults in many places (Fig. 37F). Most faults terminate at the top of the deformed bed, and many are developed entirely within the bed.

Tilted, triangular blocks of sandstone, which are bounded on one side by normal faults, are in some outcrops (Fig. 39) and

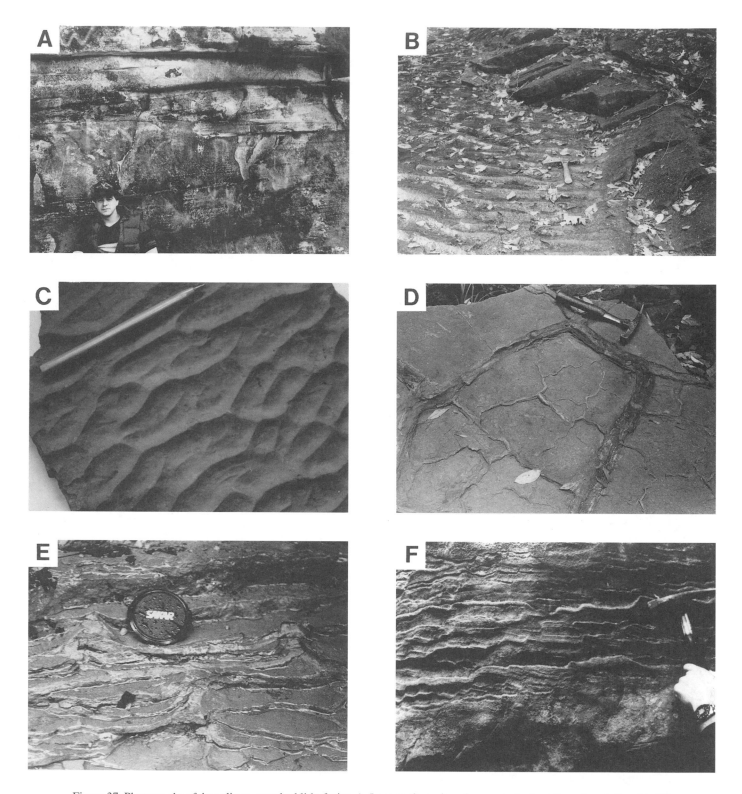

Figure 37. Photographs of the solitary-crossbed lithofacies. A, Low-angle wedge-planar crossbeds at Doan Brook (locality 33) provide evidence for local beach deposition. B, Planar-tangential crossbed extending into asymmetric, sinuous crested ripples at Stebbins Gulch (locality 42). C, Interference wave ripples from flaggy beds at Linesville, Pennsylvania (locality 58). D, Irregular, polygonal mudcracks filled with sandstone at the Bartholomew Quarry (locality 62). E, Side view of polygonal cracks at the Bartholomew Quarry. Note that sand has been inserted from below and not from above, as would be expected in desiccation structures. F, Microfaults in the lower part of the solitary-crossbed lithofacies at the Bedford Glen (locality 35).

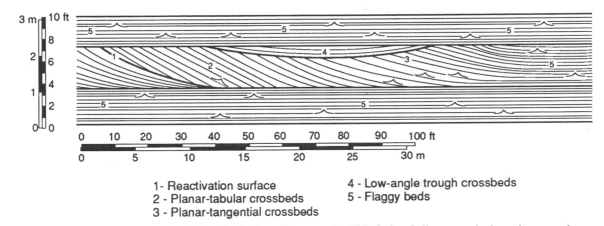

1- Reactivation surface
2- Planar-tabular crossbeds
3- Planar-tangential crossbeds
4 - Low-angle trough crossbeds
5 - Flaggy beds

Figure 38. Sketch of stratal relationships in the solitary-crossbed lithofacies. Solitary crossbeds are interpreted to represent storm-influenced tidal bars, and the flaggy beds are interpreted to represent interbar sand-mud flats.

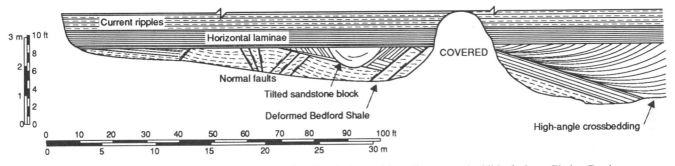

Figure 39. Tilted, fault-bound sandstone blocks forming the base of the solitary-crossbed lithofacies at Phelps Creek (locality 49). The fault blocks, as well as the top of the Bedford Shale, are truncated below horizontally laminated sandstone, indicating that significant disconformities exist at the base of and within the solitary-crossbed lithofacies.

are as much as 7 m (23 ft) thick. At Phelps Creek (locality 49), tilted blocks are exposed just downstream from a thrust fault in the Cleveland Shale, which was documented by Potter et al. (1983). Deformed shale masses more than 3 m (10 ft) thick are present locally at the top of the Bedford. At Euclid Creek (locality 34; Fig. 40), sandstone with wedge-shaped crossbeds is adjacent to one of the masses. Crossbed dip decreases upward through the unit, and adjacent to the deformed shale, crossbeds are locally oversteepened. The largest deformation structure is exposed in the eastern part of the outcrop at Granger Road (locality 32; Fig. 41), Here, the lower part of the solitary-crossbed lithofacies is more than 10 m (33 ft) thick and 30 m (100 ft) wide and is faulted and convoluted; it is bounded on the bottom and sides by complexly folded and faulted red shale resembling that associated with the deformed silty-sandstone lithofacies.

Silty sandstone and shale forms the top of the solitary-crossbed lithofacies in most exposures in northeastern Ohio (Fig. 34). These strata tend to overlie gradationally the main part of the solitary-crossbed lithofacies and generally form a coarsening-upward sequence that caps the lithofacies. The silty sandstone typically contains poorly oriented wave ripples with longer wavelength (3-7 cm; 1–3 in) than in other parts of the facies. The rippled strata are in laminae to thick beds and commonly form shaly intervals with wavy, lenticular, and flaser bedding; feeding burrows and trails are locally abundant. A thick bed of silty sandstone that ranges in thickness from 30 cm (12 in) to 150 cm (60 in) commonly forms the top of the Berea. In most places, the bed contains hummocky strata or horizontal laminae. At some locations, the unit contains scour-and-fill structures or a solitary, tangential crossbed with heterolithic toesets that extend downward into the shaly beds. Poorly preserved bivalves, tentatively identified as *Palaeoneilo* sp., and the inarticulate brachiopod, *Lingula* sp., are present locally near the top of the silty sandstone.

Crossbed azimuths in the solitary-crossbed lithofacies are variable compared to other directional structures in the Bedford-Berea sequence (Lewis, 1988); vector magnitude varies from 0.42–0.83 where more than five readings were taken. Variation is local, however, and most vector-mean azimuths in northern Ohio are northeast and range from 328°–81° (Fig. 16). Vector azimuths in northwestern Pennsylvania are even more variable and range from 322°–142°. Where five or more wave-ripple azimuths were measured, vector magnitudes were low, ranging from 0.64–0.82.

The solitary-crossbed lithofacies has a limited subsurface

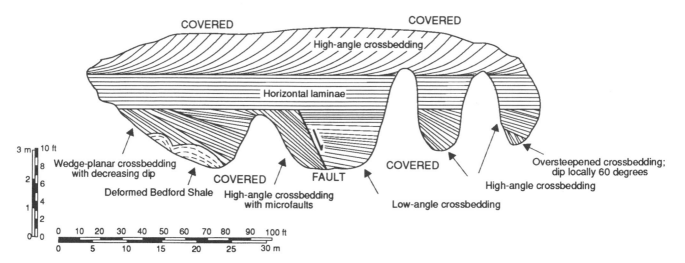

Figure 40. Deformation structures at the base of the solitary-crossbed lithofacies at Euclid Creek (locality 34) include oversteepened crossbedding and wedge-planar crossbed sets bound by deformed shale masses.

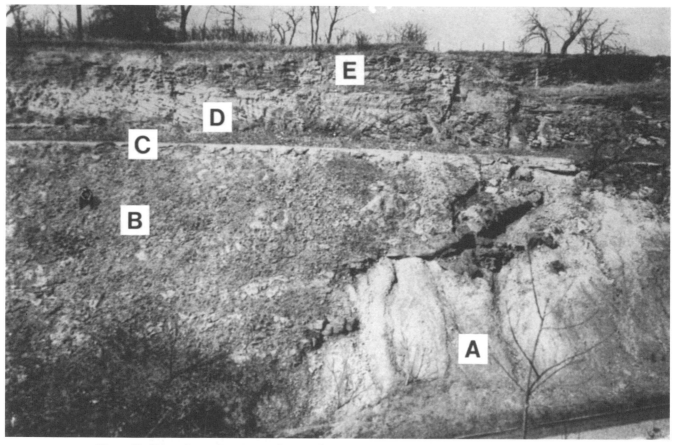

Figure 41. Although now overgrown, the most famous exposure of the solitary-crossbed lithofacies is at Granger Road (locality 32), where deformed sandstone overlies red shale and is truncated below a large-scale crossbed (after Pepper et al., 1954). A, Deformed red shale. B, Deformed sandstone at base of solitary-crossbed lithofacies; note localized coherent bedding near margin of body. C, Truncation surface at top of deformed sandstone. D, Large solitary crossbed. E, Silty sandstone with crossbeds and wave ripples.

distribution. In the subsurface of Cuyahoga and Geauga Counties, northeastern Ohio, the solitary-crossbed lithofacies is fairly uniform in thickness, which is typically between 13–20 m (40–60 ft) (Fig. 28). The sandstone passes into siltstone approximately 25 mi (42 km) south of the northernmost outcrops and thins eastward into western Ashtabula and Trumbull Counties. Because the solitary-crossbed lithofacies is in contact with the Cussewago (pebbly-sandstone lithofacies) in the subsurface, the solitary-crossbed lithofacies was mapped with the Cussewago–Second Berea Sandstone in northeasternmost Ohio and northwestern Pennsylvania; only the siltstone forming the top of the Bedford-Berea sequence was mapped as Berea in this area (Figs. 28, 34).

Environmental interpretation. The solitary-crossbed lithofacies was derived mainly by reworking Cussewago–Second Berea fluvial and deltaic deposits. The lithofacies signifies destruction of the Cussewago fluvial-deltaic system and the development of an estuarine embayment in northeastern Ohio and northwestern Pennsylvania. In the following discussion, the lithofacies is interpreted to represent an estuarine bar belt with localized beaches and sandwaves. As the estuarine embayment matured, subtidal, storm-dominated sedimentation became widespread.

Significant erosional discontinuites exist within the solitary-crossbed lithofacies, and the basal contact appears disconformable (Figs. 7, 34). This point is best illustrated at Phelps Creek (locality 49), because sandstone truncates not only faulted blocks within the lithofacies, but also the upper part of the Bedford Formation (Fig. 39). One implication of this disconformity is that Cussewago strata may have extended farther west than is now apparent. An erosional contact is also apparent in northwestern Pennsylvania, where the solitary-crossbed lithofacies overlies the Riceville Shale. The compositional similarity of sandstone in the solitary-crossbed and pebbly-sandstone lithofacies supports that the solitary-crossbed lithofacies is palimpsest Cussewago sediment. The northernmost part of the Cussewago–Second Berea lobe is truncated by the solitary-crossbed lithofacies and therefore does not reflect the original geometry of the delta (Figs. 7, 12). Truncation of the Cussewago Sandstone by the solitary-crossbed lithofacies marks destruction of the Cussewago–Second Berea fluvial-deltaic system and formation of an estuarine embayment in the northeastern part of the study area.

Solitary crossbeds (Figs. 35, 38) represent migrating bars, which are common in fluvial, estuarine, and beach systems (Allen, 1963). Wave ripples, burrows, and fossils in the solitary-crossbed lithofacies suggest marine influence, and the solitary crossbeds in the Berea are most similar to those in estuarine shoals (Terwindt, 1981; Boersma and Terwindt, 1981) and in bar-and-trough (ridge-and-runnel) beaches (Davis et al., 1972; Beets et al., 1981). Wave ripples predominate in meteorologically sensitive intertidal bar-and-trough systems, and deposits similar to the solitary-crossbed lithofacies are forming today on mesotidal beaches along the North Sea Coast (van den

Berg, 1977; Beets et al., 1981). Similar ancient deposits have been described by Bigarella (1965) and Roep et al. (1979).

Crossbedding patterns in intertidal shoals are related to tidal cycles (Terwindt, 1981; Kohsiek and Terwindt, 1981), and cyclicity of lamina thickness suggestive of tides is developed locally in the Berea (Duncan and Wells, 1992). Spring tide causes development of reactivation surfaces and tabular crossbeds, whereas neap tide causes a decrease in bar amplitude; the foresets become tangential and pass laterally into horizontal, rippled beds (Fig. 38). Short ripple wavelength in the flaggy sandstone beds typifies extremely shallow water and is characteristic of beach troughs (e.g., Davis et al., 1972; Beets et al., 1981) and intertidal flats (e.g., Evans, 1965; McCave and Geiser, 1979).

The great lateral extent of the flaggy, rippled beds suggests that they are interbar sand-flat deposits, and the abundance of wave ripples suggests that wind-generated waves were important mechanisms for sedimentation on the sand flats. Wind and storms may have also influenced reactivation and migration of bars. On North Sea intertidal beaches, ephemeral swash zones develop on some bar backs at neap tide (van den Berg, 1977; Beets et al., 1981). Perhaps the low-angle trough sets at the top of some solitary crossbeds at Phelps Creek (locality 49) and the Bartholomew Quarry (locality 62) (Fig. 38) represent similar swash deposits.

Dissipative beaches have numerous bars that dampen most wave energy, whereas reflective beaches have few bars and steep, linear beach faces that reflect most energy (Wright et al., 1979). Thus, the solitary-crossbed lithofacies represents an extremely dissipative shore-zone system, but low-angle, wedge-planar crossbeds (Fig. 37A) signify local development of reflective surf and swash zones. On intertidal beaches, bar slipfaces scarcely exceed 2 m (6 ft) in height (e.g., Hine, 1979), but some crossbeds in the solitary-crossbed lithofacies are thicker than 6 m (19 ft) (Fig. 36). These crossbeds commonly contain clay drapes and sigmoidal topsets, which are associated with tidal processes (Terwindt, 1981; Kreisa and Moiola, 1986). Similar giant crossbeds have been interpreted to be subtidal sand-wave deposits by Nio (1976) and Allen (1982). Low crossbed and wave-ripple consistency suggests that the bar belt was poorly organized. However, northeast vector-mean crossbed azimuths in northeastern Ohio (Fig. 16) establish a northeast flooding direction. In northwestern Pennsylvania, poor alignment of bedforms may be related to topography inherited from Cussewago erosion and deposition.

Deformation in the solitary-crossbed lithofacies is clearly synsedimentary, because faults are truncated by younger strata (Coogan et al., 1981; Lewis, 1988; Duncan and Wells, 1992) (figs. 39, 40). Much of the deformation is typical of structures formed by loading of sand on mud (McKee and Goldberg, 1969; Parker, 1973). En echelon faulting at Phelps Creek (Fig. 39) is suggestive of slumps such as those documented in stream-bank failures by Laury (1971) and may be related to thrusting in the Cleveland Shale. Strata are contorted in many places where the solitary-crossbed lithofacies overlies more than 5 m (16 ft) of red shale, and contortion of part of the sandstone body at Granger Road (locality 32; Fig. 41) suggests that sediment was fluidized—sand apparently collapsed into red mud at the east end of the outcrop. This structure contrasts strongly with the faulted structures farther east, illustrating the incompetence of clay-rich mud relative to silty sediment.

The fauna (*Palaeoneilo*, *Lingula*, *Bifungites*) and sedimentary structures in the silty sandstone at the top of the lithofacies (Fig. 34) indicate subtidal sedimentation, and abundant wave ripples and hummocky strata indicate storm action below fair-weather wave base (Dott and Bourgeois, 1983; Swift et al., 1983). Hence, this part of the lithofacies represents deepening and maturation of the estuarine embayment. The general coarsening-upward sequence in this part of the lithofacies indicates progradation, but the silty sand apparently was not derived entirely from the underlying solitary-crossbedded units because silty sandstone is only locally in contact with sandstone and is much finer grained. Similar silty sandstone above the quarrystone lithofacies indicates that some sediment may have been derived by reworking the Cussewago delta front (Fig. 7). Extension of the silty sandstone into the Berea Siltstone of northwestern Pennsylvania (Figs. 7, 34) indicates that reworking of the delta front provided a source of silt and sand that prograded eastward into the estuarine embayment.

Siltstone lithofacies

Characteristics. The siltstone lithofacies contains at least 33% medium- and thick-bedded siltstone and is one of the most widespread facies of the Bedford-Berea sequence. The lithofacies comprises the Euclid and Sagamore Members of the Bedford Formation, the Cussewago–Second Berea siltstone belt, and the Berea Siltstone (Figs. 7–10). Siltstone beds are separated into two types, unrippled (those lacking wave ripples) and rippled (those containing wave ripples) (Fig. 42).

Unrippled siltstone beds (Fig. 42) are planar and have sharp basal contacts with feeding burrow casts, groove casts, and isolated shale chips. Deformational features include load casts and ball-and-pillow structures. Upper contacts tend to be gradational and burrowed (Fig. 43A). These beds ideally contain complete Bouma (1962) sequences in which convolute laminae are present in lieu of current-ripple cross-laminae in division T_c (Fig. 43B). Although many of these sequences are amalgamated, unrippled siltstone beds are separated by thinly interbedded siltstone and shale with wavy and lenticular bedding; these beds typically contain partial Bouma T_{cde} sequences with current-ripple cross-laminae.

Rippled siltstone beds (Fig. 42) have sharp, planar to wavy basal contacts with feeding-burrow casts, prod marks (Fig. 43C), and groove casts. Like unrippled siltstone, load casts and ball-and-pillow structures are common. Upper contacts universally contain straight-crested wave ripples (Fig. 43D) and are sharp, planar to wavy, and contain feeding bur-

rows, trails, and resting traces. Thick siltstone beds ideally contain complete hummocky sequences (cf. Dott and Bourgeois, 1982) (Fig. 42). Although hummocky strata are locally developed (Fig. 44A), many of the thickest beds contain swaley strata (Fig. 44B). Beds with swaley strata are wavy or broadly lensoid, and wave ripples are superimposed on many of the swales. As in unrippled siltstone, many of these sequences are amalgamated, but most unrippled siltstone beds are separated by thinly interbedded siltstone and shale with wavy, lenticular, and flaser bedding. Thin siltstone beds typically contain partial hummocky sequences.

Hyde (1911) discovered the consistent northwest orientation of Bedford-Berea wave ripples in central and southern Ohio, and Lewis (1968, 1988) demonstrated that this configuration extends into northeastern Ohio. Indeed, vector-mean azimuths from the siltstone lithofacies and the rippled beds of the Bedford (gray shale lithofacies) range from 284° to 316° in most of the state (Fig. 45). The vector magnitudes in rippled siltstone exceed 0.90 in most of the study area. Wave ripples in the Bedford Shale and Berea Siltstone of extreme northeastern Ohio and northwestern Pennsylvania, however, are less consistent, and vector magnitudes are as low as 0.67. The ripple crests strike north-south (336°–35°) and have nearly twice the range of all the other wave-ripple azimuths in the siltstone lithofacies. Only in this area are any wave ripples in siltstone oriented northeast.

Euclid and Sagamore Members. The Euclid Member is a northeast-trending siltstone body within the Bedford Shale of Cuyahoga County, northern Ohio, that rises northward in section relative to the Cleveland Shale and thickens from 4–10 m (15–33 ft) (Fig. 46). The Euclid is present where gray Chagrin Shale and siltstone passes westward into black and gray Olmsted Shale and also where gray shale passes westward into red shale in the Bedford (Fig. 7). In the type section at Euclid Creek (locality 34), the Euclid Member contains exclusively rippled siltstone beds, whereas farther south at Granger Road (locality 32) and Doan Brook (locality 33), unrippled siltstone beds predominate in the lower part of the member (Fig. 46).

At Quarry Creek (locality 30), west of the main Euclid siltstone trend, approximately 6.5 m (21 ft) of deformed black and gray silty shale and gray siltstone is at the base of the Bedford (Fig. 46). Here, the base of the Bedford is marked by two graded siltstone beds. Above these beds, black silty shale is contorted and contains folded siltstone dikes (Fig. 43E) with stretched load casts. The shale contains lensoid masses of convoluted siltstone as much as 1.5 m (4.5 ft) thick and 5 m (16 ft) wide; the masses have convex bottoms and planar tops (Fig. 43F).

The Sagamore Member, which ranges in thickness from 3–7 m (10–23 ft), is the largest of several bundles of rippled siltstone beds in the Bedford of Cuyahoga and Summit Counties (Fig. 47). The Sagamore is lithologically similar to the Euclid, although the Sagamore contains a larger proportion of shaly strata and lacks unrippled beds. Siltstone bundles similar to the Euclid and Sagamore Members are present at several

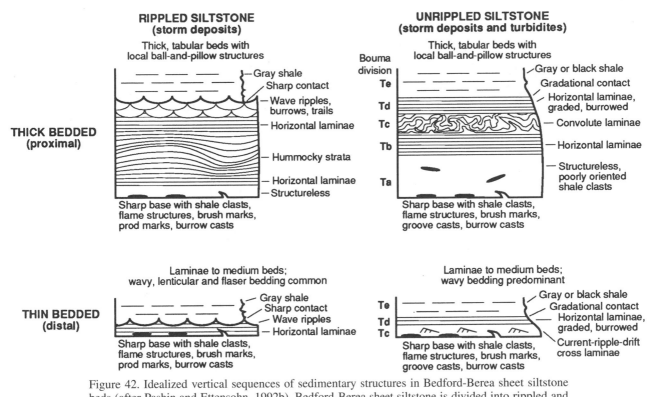

Figure 42. Idealized vertical sequences of sedimentary structures in Bedford-Berea sheet siltstone beds (after Pashin and Ettensohn, 1992b). Bedford-Berea sheet siltstone is divided into rippled and unrippled types. Rippled sheet siltstone is interpreted to represent storm-dominated shelf deposits, whereas unrippled sheet siltstone is interpreted to represent turbidite apron or fan deposits as well as combined-flow storm deposits. Thick sheet-siltstone beds are proximal storm deposits and turbidites, and thin beds are distal examples.

localities but are too thin and too local to be included in the siltstone lithofacies. These bundles, like the Sagamore Member, are generally present farther east and higher in section than the Euclid Member.

Cussewago–Second Berea siltstone belt. The Cussewago–Second Berea siltstone belt extends southward from Tuscarawas County into Gallia County (Figs. 12, 17). The belt narrows southward from 40–16 km (25–10 mi). The siltstone is locally thicker than 8 m (25 ft) and is thickest near its eastern margin. The western margin of the belt is irregular and approximately parallels the platform margin from Muskingum County south (Figs. 11, 12, 17). The eastern margin of the siltstone belt is more regular than the western margin, and localized patches of siltstone at the level of the Cussewago–Second Berea are mainly east of the belt in Belmont and Monroe Counties.

Berea Siltstone. The Berea Siltstone is the most widespread part of the siltstone lithofacies, extending through most of the study area (Figs. 34, 48). Wavy, swaley-stratified siltstone beds (Fig. 44B) predominate in the outcrops of Ohio and northwestern Pennsylvania (Figs. 34, 48). In southern Ohio, the Berea Siltstone thickens southward and forms the "Cliff Stone" of Hyde (1953); these strata are typically planar and lack swaley strata. The Berea siltstone typically is a thickening- and coarsening-upward sequence (Fig. 48), but in northwestern

Pennsylvania, the siltstone forms a thinning- and fining-upward sequence that grades into Orangeville Shale (Fig. 34).

In the subsurface of northern Ohio, the Berea Siltstone is contiguous with the Berea Sandstone and is separated from the Cussewago–Second Berea Sandstone by Bedford Shale (Figs. 8–10, 33). In central Ohio, a lobate body of siltstone and sandstone outlined by the 25-ft (8-m) contour extends southward into Delaware and Franklin Counties from the area containing the quarrystone, deformed silty sandstone, and low-angle cross-bed lithofacies (Figs. 28, 33). An isolated area with siltstone thicker than 8 m (25 ft) is in south-central Ohio; this part of the siltstone lithofacies straddles the platform margin and is built around the thick, ostensibly deformed sandstone bodies of Hocking County (Figs. 10, 28).

On the eastern platform of northeastern and east-central Ohio, Berea Siltstone is typically 8–18 m (25–60 ft) thick, forming a widespread blanket that extends as far south as Morgan County above the Cussewago–Second Berea siltstone belt (Figs. 17, 33). This part of the lithofacies is contiguous with the silty sandstone of the solitary-crossbed lithofacies and the Berea Siltstone in the outcrops of northeastern Ohio and northwestern Pennsylvania. In northwestern Pennsylvania, the Berea Siltstone passes into the shale of the Shellhammer Hollow Formation. East of the Shellhammer Hollow Formation, the Corry

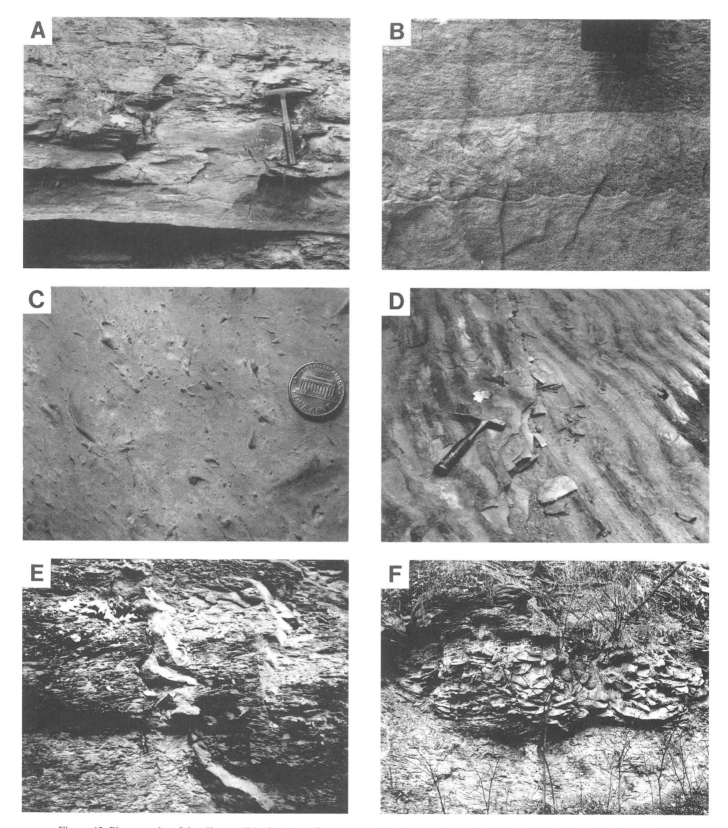

Figure 43. Photographs of the siltstone lithofacies. A, Graded siltstone bed in black silty shale at Quarry Creek (locality 30). Note sharp base and indistinct top. B, Internal convolute laminae in unrippled siltstone bed of the Euclid Member at Doan Brook (locality 33). C, Prod marks from base of rippled siltstone bed at Doan Brook. D, Wave ripples on top of siltstone bed in the Sagamore Member at Bridal Veil Falls (locality 36). E, Ptygmatically folded siltstone dike in black silty shale at Quarry Creek (locality 30). F, Lensoid ball-and-pillow mass in black silty shale at Quarry Creek. Note planar top and convex base; bed is 0.9 m (3 ft) thick.

Figure 44. Hummocky and swaley strata in rippled siltstone beds of the siltstone lithofacies. A, Hummocky strata at Stratton Creek (locality 57). Bed is 12 cm (0.4 ft) thick. Hummocky strata in the Bedford-Berea sequence are generally preserved as complete bedforms and are rarely cross stratified. B, Swaley strata at Gahanna (locality 10). Swaley strata are abundant in the Berea Siltstone and apparently represent preferential preservation of swales over hummocks as storms waned. At Gahanna, swaley strata include compound bedforms containing superimposed wave ripples.

Sandstone is present where the Cussewago Sandstone is absent (Figs. 12, 28). In several areas underlain by thick Cussewago–Second Berea Sandstone the Berea siltstone blanket is thicker than 15 m (50 ft) (Figs. 11, 28). Along the platform margin, thickness of the Berea varies considerably, ranging from 0–3 m (0–10 ft) in part of Summit County to more than 23 m (75 ft) in eastern Wayne and north-central Holmes Counties (Fig. 28).

In Scioto and Lawrence Counties, south-central Ohio, the Berea Siltstone thickens southward at the expense of the Bedford Shale (Figs. 7, 8, 28). The siltstone has a maximum thickness of 48 m (140 ft) and crops out as the Cliff Stone of Hyde (1953) in south-central Ohio and northeastern Kentucky (Fig. 49). This part of the Berea is the northern part of the extensive Berea Siltstone of Kentucky and western West Virginia (Fig. 20). Along the West Virginia–Kentucky border, the siltstone thickens westward from approximately 12 m (40 ft) to more than 30 m (100 ft). This configuration reflects concomitant thickening of the Bedford-Berea sequence and southward extension of the eastern platform and the western basin into West Virginia and Kentucky (Figs. 20, 49). In the basin the Berea thins westward and intertongues with Bedford Shale.

Environmental interpretation. The siltstone lithofacies is interpreted to represent a spectrum of storm-related turbidite and shelf deposits that reflect progressive shoaling of the western Appalachian basin. The Euclid Member and the Cussewago–Second Berea siltstone belt are localized, shelf-margin silt accumulations that formed early in Bedford-Berea deposition. The Berea Siltstone of northern Ohio evidently represents regional shoaling during the late stages of delta destruction and formation of a storm-dominated silt blanket that prograded into the estuarine embayment. By contrast, the Berea Siltstone of south-central Ohio, eastern Kentucky, and West Virginia is a thick package of storm deposits and turbidites that prograded into the western basin from fluvial-deltaic sources associated with the Gay-Fink and Cabin Creek trends.

Abundant wave-related bedforms in the rippled beds (Fig. 42), such as wave ripples (Fig. 43D) and hummocky strata (Fig. 44A), are characteristic of storm-dominated shelf deposits, which form in the transition zone between fair-weather and storm wave base (e.g., Goldring and Bridges, 1973; Dott and Bourgeois, 1982). In the North Sea, which is one of the most viable analogs for ancient epeiric seas, fair-

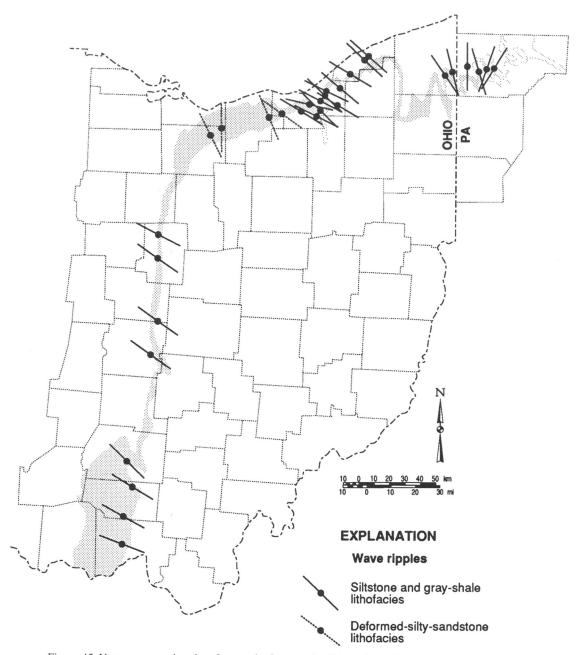

Figure 45. Vector-mean azimuths of wave-ripple crests in Ohio and northwestern Pennsylvania.

weather wave base is between 10–30 m (33–100 ft) below sea level, and storm wave base is approximately 70 m (210 ft) below sea level (e.g., Hamblin and Walker, 1979; Aigner and Reineck, 1982). Shelf storm deposits are now routinely recognized in ancient successions (e.g., Hamblin and Walker, 1979; Bourgeois, 1980), including the Catskill clastic wedge (Craft and Bridge, 1987), and a storm-dominated shelf origin has been proposed for rippled beds in the Bedford-Berea sequence by many investigators (Rothman, 1978; Potter et al., 1983; Pashin and Ettensohn, 1987; Lewis, 1988).

Unrippled siltstone beds commonly contain complete Bouma sequences, which are typical of turbidites (Fig. 42).

Storm deposits and turbidites are closely related in many depositional sequences, thus storm-generated turbidite deposition may have been common in the geologic past (Hamblin and Walker, 1979; Bourgeois, 1980; Woodrow and Isley, 1983). The thick-bedded Bedford-Berea turbidites are lithologically similar to those identified in prodeltaic aprons of the Pocono wedge (Moore and Clarke, 1970; Kepferle, 1977, 1978; Chaplin, 1980), but Pashin and Ettensohn (1987) interpreted Bedford-Berea examples in northeastern Kentucky and south-central Ohio to have been generated by storms in nearby shelf environments. Hence, some unrippled beds may simply represent silt that was transported by storms into areas where the substrate

EXPLANATION

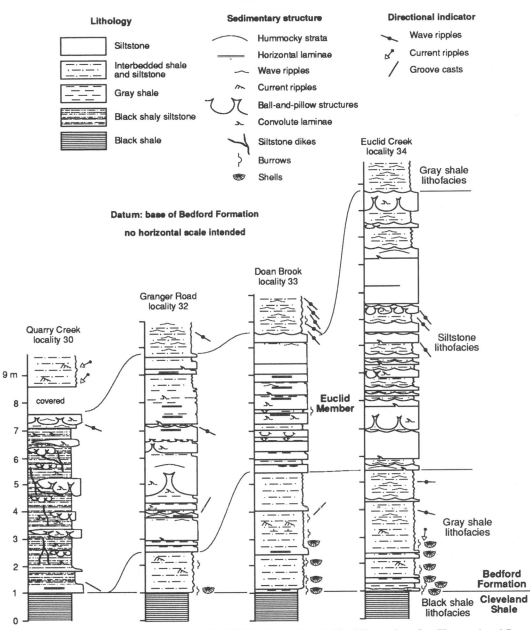

Figure 46. Measured sections of the Euclid Member of the Bedford Formation. See Figures 4 and 7 for location.

was below storm wave base, and thus, where gravitational density flows could operate. Beds resembling turbidites may also represent storm-generated combined-flow deposits (Nelson, 1982; Myrow and Southard, 1991) and, as is discussed in the following sections, unrippled siltstone beds in the Bedford-Berea sequence apparently have diverse origins.

Euclid and Sagamore Members. The Euclid Member contains a distinctive suite of storm deposits, including combined-flow deposits that superficially resemble turbidites. The member contains more rippled siltstone beds in the north than in the south, so facies gradient and west-northwest wave-ripple orientation establish that the member was deposited on a southwest

slope. Lewis (1988) recognized that presence of the Euclid in a linear body above the Chagrin-Olmsted transition suggests that it was deposited by shoaling along a shelf margin formed by relict topography and differential compaction of black and gray mud (Fig. 50). The presence of both rippled and unrippled siltstone beds in the Euclid Member suggests that the unrippled beds were transported along the shelf margin by combined unidirectional and oscillatory flows.

The planar, graded siltstone beds in black shale (Fig. 43A) at Quarry Creek (locality 30; Fig. 46) are suggestive of turbidites, but association with clastic dikes (Fig. 43E), ball-and-pillow lenses (Fig. 43F), and deformed black shale is unusual.

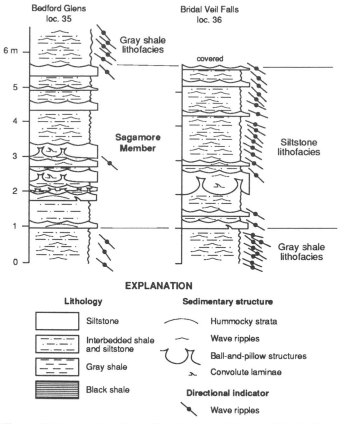

Figure 47. Measured sections of the Sagamore Member of the Bedford Formation. See Figures 4 and 7 for location.

Surlyk (1987) described similar features in Jurassic black shale in Greenland. He interpreted deformed sandstone lenses to be turbiditic gully fills and much of the deformation and dike injection to be related to postburial compaction and seismicity. The gullies described by Surlyk formed along fault-controlled slopes. By contrast, the Bedford examples apparently represent Euclid sediment that was transported across the Chagrin-Olmsted shelf margin onto a fluid substrate (Lewis, 1988). The turbiditic gully fills are preserved with the only Bedford-Berea black shale in Ohio, so one interpretation is that silt and mud were transported into a depression west of the slope break that restricted circulation and enhanced preservation of organic matter.

The Sagamore Member, like other siltstone bundles in Cuyahoga and Summit Counties, is composed of rippled storm deposits (Figs. 42, 47). The local nature of these bodies suggests deposition in open-shelf siltstone patches. Shelf sand patches are thought to form by bypass of sediment from nearshore environments into shoal areas (Brenner, 1978). The Sagamore Member and the other siltstone bundles in the Bedford may also have formed in this manner. The presence of Sagamore-type siltstone bundles progressively farther eastward and higher in section suggests shoaling at a shelf break that retreated eastward.

Cussewago–Second Berea siltstone belt. In Tuscarawas County, the Cussewago–Second Berea Sandstone passes into the siltstone belt (Figs. 12, 17), which was interpreted by Pepper et al. (1954) to be a barrier island complex. However, the fine grain size of the belt suggests that it includes an open-marine silt body that perhaps resembled the Euclid and Sagamore Members. In Muskingum, Perry, and Athens Counties, the position of the siltstone belt along the platform margin suggests deposition by southward storm-related flows that hugged the basin margin and caused shoaling of sediment derived from the Cussewago delta and its delta-destructive beaches. The undulatory western margin of the belt (Fig. 17) is suggestive of fanlike bodies and may represent silt that was transported westward across the platform margin. The area east of the siltstone belt is interpreted as a shelf or lagoon containing isolated silt patches.

Berea Siltstone. Although storm deposits in the Bedford are simple sheet-siltstone beds, undulatory to lensoid siltstone beds with swaley strata (Fig. 44B) abound in the Berea (Figs. 34, 48). Swaley strata apparently form in shallower, higher energy environments than hummocky strata, commonly above fair-weather wave base (Leckie and Walker, 1982; Cotter, 1985; McCrory and Walker, 1986). Berea examples are capped by shale, however, indicating deposition below fair-weather wave base. Swaley strata in the Berea apparently represent preferential preservation of swales by reworking hummocks and smoothing bedtops as storms waned.

Wave-ripple crests in shelf environments generally parallel depositional strike (Vause, 1959), which was therefore west-northwest in most of the study area (Fig. 45). Marginal-marine sandstone passes southward into open-marine siltstone, thus regional paleoslope was southwest, and the high consistency of wave-ripple orientation in this area suggests that topographic irregularity was minimal. The variability of wave-ripple orientation in northwestern Pennsylvania is interpreted to reflect topographic irregularity that remained following formation of the estuarine embayment. For example, the presence of the Corry Sandstone where the Cussewago Sandstone is absent (Figs. 12, 28) may signify shoaling on a submerged erosional high between Cussewago channel axes.

In north-central Ohio, the Berea Siltstone is in contact with the Berea Sandstone and not with the Cussewago–Second Berea Sandstone (Figs. 7–10, 34). This means that the Cussewago delta had been inundated fully before siltstone deposition and that the siltstone was derived at least partly by reworking the Cussewago delta front. Hence, some storm-reworked deltaic sediment prograded back toward the sediment source as the estuarine embayment subsided. This reworking apparently affected even the distalmost parts of the delta front, as indicated by the isolated blanket of siltstone centered around the thick, deformed sandstone in Hocking County. Thickening and coarsening of the siltstone blanket above the Cussewago–Second Berea in east-central Ohio, moreover, suggests shoaling on relict topographic highs formed by the abandoned delta. Pronounced irregularities of thickness along the platform margin, moreover, suggest that it still had significant topographic relief.

The Berea Siltstone of eastern Kentucky and south-central

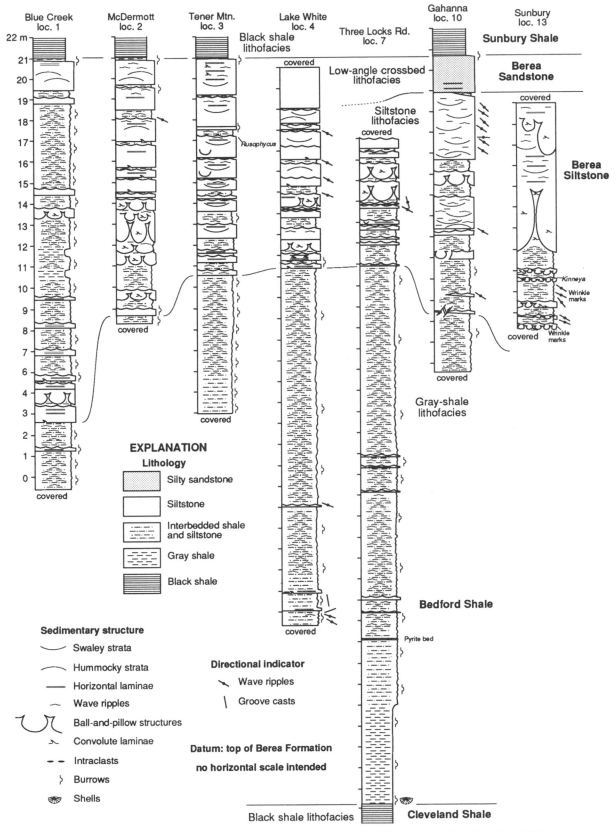

Figure 48. Measured sections of the Berea Siltstone and Bedford Shale in central and southern Ohio. See Figures 4 and 7 for location.

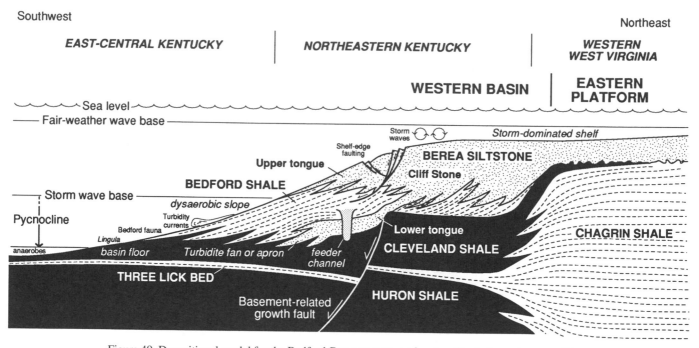

Figure 49. Depositional model for the Bedford-Berea sequence of eastern Kentucky and western West Virginia (after Pashin and Ettensohn, 1992b). A thin, aggradational shelf sequence was preserved on the eastern platform, whereas a thick, progradational sequence that spans shelf, slope, and basinal environments was preserved in the western basin. Differentiation of platform and basin areas was related to relict topography and differential compaction of organic-rich black mud (Cleveland Shale) and relatively incompactible, organic-poor gray mud and silt (Chagrin Shale). Within the platform and basin areas, however, basement-fault reactivation was a significant control on facies distribution.

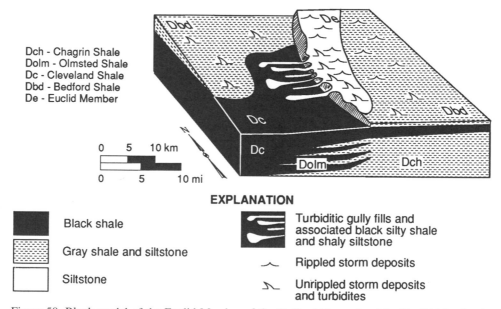

Figure 50. Block model of the Euclid Member of the Bedford Formation. The Euclid Member is interpreted to represent shelf-margin storm deposits and associated turbidites that accumulated where gray Chagrin Shale passed westward into interbedded gray and black Olmsted Shale.

Ohio represents a westward-prograding wedge of shelf deposits and slope turbidites (Pashin and Ettensohn, 1987) (Fig. 49). The location of the siltstone wedge in eastern Kentucky indicates that it was derived from the Gay-Fink and Cabin Creek trends of West Virginia (Fig. 20). Southward deepening of the shelf was controlled by the Rome trough basement faults, as indicated by the localized parallelism of the isopach contours to the faults.

Gray shale lithofacies

Characteristics. The gray shale lithofacies is that part of the Bedford-Berea sequence containing more than 67% gray shale and thinly bedded siltstone. The lithofacies is present throughout the field area and includes gray Bedford Shale and the Shellhammer Hollow Formation (Fig. 7). In Ohio, the contact between the gray shale lithofacies and the Cleveland Shale is distinct and planar. The lithofacies is overlain gradationally by the Berea Siltstone and sharply by the Berea Sandstone. In central Ohio, red Bedford Shale (red shale lithofacies) forms a northward-thickening wedge between strata of the gray shale lithofacies. In northern Ohio, however, an eastward-thinning wedge of the red shale lithofacies separates the gray shale lithofacies from the Berea Sandstone.

Gamma-ray logs demonstrate that, as in outcrop, the Bedford Shale sharply overlies the black Cleveland Shale in the subsurface (Figs. 8–10). In parts of the western basin more than 50 m (150 ft) of the Bedford separates the Berea from the Cleveland, and thickness variation is greatest amid the deformed sandstone masses of north-central Ohio. Cuttings indicate that the red and gray shale units intertongue profusely and that red shale is largely enveloped by gray shale. On the eastern platform, however, the Bedford Shale is generally thinner than 16 m (50 ft). The Bedford thins by as much as 38 m (125 ft) at the platform margin and extends between the Cussewago–Second Berea and the Berea into southeastern Ohio. Where the Cussewago–Second Berea Sandstone is present, the Bedford Shale is generally less than 8 m (25 ft) thick and is locally absent.

In outcrop the gray shale lithofacies is generally thinner than 30 m (100 ft) and contains as much as 23 m (80 ft) of rippled beds in the upper part and as much as 20 m (70 ft) of unrippled beds in the lower part (Figs. 34, 48). Rippled strata predominate in northeastern Ohio, whereas unrippled strata predominate in central Ohio where they locally constitute the whole lithofacies (Fig. 32). The shale generally has an irregular, platy fissility (Fig. 51A), except in the diapiric shale masses of north-central Ohio. Wavy and lenticular bedding predominates (Fig. 51B), and trace fossils and sedimentary structures are the same as those in thinly bedded parts of the siltstone lithofacies. Thin, rippled siltstone beds generally contain the upper parts of hummocky sequences (Fig. 42), whereas thin, unrippled beds are commonly graded and contain successions resembling Bouma T_{cde} sequences (Figs. 42, 52A) or they lack grading and

contain asymmetrical ripple bedforms (Fig. 52B). The Bedford fauna is in the lower 3 m (10 ft) of the gray shale lithofacies and is most abundant at the basal contact (Pashin and Ettensohn, 1992a).

Paleocurrent data for wave ripples in the gray-shale lithofacies are essentially the same as in the siltstone lithofacies (Figs. 16, 45). Directional data for groove casts were taken largely from unrippled siltstone beds, but a few readings were obtained from rippled beds. Groove-cast readings are variable with the vector-mean azimuth ranging from 289° to 85°; most are perpendicular to wave-ripple crests, although some are nearly parallel. Where more than one reading was taken, the vector magnitude ranges from 0.73 to 1.00. Most current-ripple azimuths were measured from the unrippled beds, and the lowest vector magnitude for a given locality was 0.98. Considerable variation exists among the localities, but all the vector-mean azimuths have a westward component, and most of the azimuths are southwest. Current-ripple vectors are typically oblique to wave-ripple crests in supradjacent strata by more than 45°, but in the few beds where wave and current ripples are present together, current-ripple foresets dip parallel to wave-ripple crests.

Locally, large-scale deformation structures are exposed in central and southern Ohio near the top of the gray shale lithofacies. At Gahanna (locality 10), a contorted body of shale and siltstone has a convex basal contact and a planar upper contact and is more than 30 m (100 ft) wide and 8 m (25 ft) deep (Cooper, 1943). The base of the body is defined by a fault, and the top is truncated by strata of the siltstone lithofacies. Along the truncation surface, thick siltstone beds contain recumbent folds with subsidiary open folds that have subvertical axial planes. Hyde (1953) photographed a similar structure in Pike County, Ohio (Fig. 51C), but the lower contact of the deformed body appears gradational. Bedding is not visible within the structure, save for convoluted siltstone balls near the base. Above the structure, bedding is coherent but sags slightly.

Environmental interpretation. The gray shale lithofacies was deposited in diverse environments. As discussed previously, the deformed gray shale masses of north-central Ohio include diapiric prodelta deposits that intruded the Cussewago–Second Berea delta front. The part of the lithofacies containing rippled siltstone beds apparently was deposited in muddy shelf and upper slope environments, whereas the part containing unrippled siltstone beds was deposited in muddy, dysaerobic, lower slope turbidite aprons. Similar gray shale and siltstone in northeasternmost Ohio and northwestern Pennsylvania, however, are interpreted to have been deposited in an estuarine embayment.

Thin bedding and predominance of wave ripples (Fig. 42) typify distal storm deposits (Aigner and Reineck, 1982). Because distal storm deposits in the Bedford-Berea are transitional between shelf deposits of the siltstone lithofacies and the shaly, unrippled phase of the gray shale lithofacies, they accumulated mainly in upper slope environments. In northeastern Ohio and northwestern Pennsylvania, however, a large part of

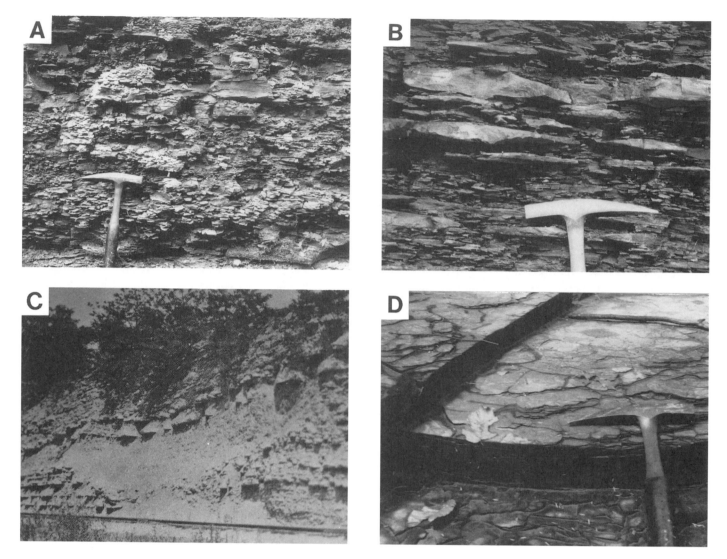

Figure 51. Photographs of the gray shale and black shale lithofacies. A, Gray shale with irregular, platy fissility in the Bedford Formation at the Bedford Glen (locality 35). B, Gray shale and rippled siltstone with wavy and lenticular bedding at the Bedford Glen. C, Slump structure near the top of the gray shale lithofacies at Waverly (after Hyde, 1953). Unfortunately, this outcrop no longer exists. D, Black Cleveland Shale with characteristic blocky jointing and well-developed platy fissility.

the lithofacies is east of the shelf-edge deposits of the Euclid Member and the fluvial-deltaic deposits of the Cussewago– Second Berea Sandstone and is therefore interpreted to represent muddy shelf and estuarine bay deposits, respectively.

The deformed shale bodies in central and southern Ohio are typical of submarine slump deposits. Presence of slump structures at the contact between the gray shale and siltstone lithofacies indicates that the Bedford-Berea shelf break was in places unstable. Submarine slumps may develop on slopes as gentle as 0.5° (Shepard, 1955) and commonly form in response to seismicity or storm-wave loading (Schwab and Lee, 1988). Recumbent folds with subsidiary, subvertical fold axes, such as those at Gahanna, indicate significant lateral transport (Farrell and Eaton, 1987). In contrast, the gradational lower contact of

the body near Waverly (Fig. 51C) suggests that lateral transport was minimal in some of these structures.

As in the siltstone lithofacies, thinly bedded, unrippled silt-stone in the gray shale lithofacies contains dominantly Bouma T_{cde} sequences (Figs. 42, 52A). The high consistency of current-ripple cross-laminae indicates that the flows were unidirectional, and the groove-cast and current-ripple azimuths, which are directed at high angles to the wave-ripple crests higher in section (Figs. 16, 45), indicate a southwestward paleoslope. Strata similar to those in the gray shale lithofacies are typically deposited in the distal parts of modern turbidite complexes (Walker, 1978; Nelson et al., 1978) and are widespread in the Catskill delta complex (McIver, 1970; Broadhead et al., 1982; Woodrow and Isley, 1983).

Figure 52. Siltstone beds in the lower part of the gray shale lithofacies contain distinctive sedimentary structures. A, Thin, graded siltstone bed from Doan Brook (locality 33) containing partial Bouma T_{cde} sequence. These beds resemble turbidites but may also represent combined-flow storm deposits. B, Asymmetrical ripple bedform from Rocky River Reservation (locality 27). The low amplitude (1 cm) and long wavelength (15 cm) of these bedforms suggest affinity with hummocky strata and, hence, deposition by combined unidirectional and oscillatory flows.

As with thick siltstone beds, however, these deposits may represent combined flows generated by storms. For example, graded siltstone beds with groove casts and current ripples directed parallel or obliquely to the wave-ripple crests higher in section are probably the products of combined oscillatory and unidirectional flows rather than pure turbidity currents. This is especially apparent in the Bedford of northern Ohio, where Lewis (1988) demonstrated that unidirectional current indicators within the rippled beds turn parallel to depositional strike upward in section. Moreover, thin siltstone beds lacking grading and containing asymmetrical ripple bedforms with low amplitude and long wavelength (Fig. 52B) may be interpreted as translational hummock-style structures formed by combined flows (cf. Nøttvedt and Kreisa, 1987).

Lundegard et al. (1985) found that Catskill turbidites were deposited in areally extensive aprons rather than in channelized fans, the thin-bedded examples accumulating mainly in lower slope environments. Submarine slumps can generate muddy turbidity currents (Morgenstern, 1967; Stow and Piper, 1984), but slumps in the Bedford-Berea sequence are localized. In the Bedford-Berea sequence, unrippled siltstone beds are everywhere overlain by storm deposits, suggesting a much stronger association with storm-wave action than with slumping. For this reason, most of these strata are interpreted to represent a series of base-of-slope aprons that formed by turbidity currents and combined flows generated as storms transported mud and silt into regions where the sea floor was below effective storm wave base.

Unrippled strata of the gray shale lithofacies are typical of

sediment deposited in the dysaerobic zone of Byers (1977), owing to the presence of the thin-shelled, brachiopod and mollusc-dominated Bedford fauna (Pashin and Ettensohn, 1992a). Identifying the top of the dysaerobic zone is difficult because fossils are scarce in most parts of the Bedford-Berea sequence. Wave ripples in the upper part of the gray shale lithofacies provide the best evidence for aerobic conditions because the water was at times vigorously circulated by storms.

Most turbidites and storm deposits apparently were deposited in the western basin where the Bedford Shale is thickest (Figs. 8–10), and the coarsening-upward sequence in the basin indicates that the shelf prograded and became regionally extensive by the end of Bedford-Berea deposition. Most of the gray shale lithofacies on the eastern platform probably represents shallow-water deposits, considering proximity to the fluvial-deltaic deposits of the Cussewago–Second Berea. However, lack of oscillatory bedforms in gray shale immediately above the solitary-crossbed lithofacies in northeastern Ohio and northwestern Pennsylvania (Fig. 7) suggests that the deepest parts of the estuarine embayment were deep enough to be protected from direct storm-wave action.

Red shale lithofacies

Characteristics. The red shale lithofacies is the part of the Bedford Shale containing at least 50% red or brown shale. In outcrop the lithofacies is restricted to northern and central Ohio, but in the subsurface it extends southward into northeasternmost Kentucky (Fig. 53). The lithofacies has a maximum thickness in outcrop of 20 m (70 ft) at Big Creek (locality 29) and thins eastward to a feather edge (Fig. 7). In central Ohio, the red shale is enveloped by the gray shale lithofacies, whereas in northeastern Ohio, it forms the top of the Bedford. In the diapiric shale masses of northern Ohio, the shale is intensely deformed and is discontinuous. Where overlain by the Berea, as much as 2.5 m (8 ft) of gray shale is included in the red shale lithofacies; it gradationally overlies the red shale and is abruptly overlain by the Berea, the basal few centimeters of which is cemented by marcasite. Most important, the gray shale maintains continuity without regard to faults and scour-and-fill structures.

Red shale generally forms thick beds containing scarce gray shale laminae and intertongues profusely with the enveloping strata of the gray shale lithofacies (Figs. 7–10). Where these strata intertongue, red shale tends to be present in thin beds separated by laminae and very thin beds of gray siltstone that are graded or horizontally laminated. Each siltstone layer is overlain by a lamina of gray shale with a gradational upper contact. Locally, red shale is interbedded with gray shale and siltstone containing wave ripples resembling that in the gray shale lithofacies. Cobble-size spheroids of gray shale containing granule- to pebble-size grains of organic matter in the center are present in some outcrops. Other than these organic particles, the red shale lacks biogenic structures.

The red shale is locally thicker than 45 m (150 ft) in the

western basin, and thickness variation is greatest in north-central Ohio (Fig. 53), reflecting extensive soft-sediment deformation. The main axis of red shale thicker than 15 m (50 ft) extends from Cuyahoga County in north-central Ohio to Jackson County in south-central Ohio. In north-central Ohio, a secondary axis of thick red shale extends from Erie County to Richland County. In general, thickening of the red shale corresponds with thickening of the Bedford-Berea sequence. Red shale is absent in the outcrops of south-central Ohio and is scarcely thicker than 23 m (75 ft) in the outcrop belt stretching from north-central to central Ohio. Cutting sets establish that the red shale intertongues with

the Berea Siltstone in Lawrence County, south-central Ohio (Fig. 8). On the eastern platform of southern Ohio, red shale generally is thinner than 15 m (50 ft) (Figs. 8–10, 53). Here, the shale reaches its eastern limit above the Cussewago–Second Berea Sandstone along a line extending northeastward from Gallia County to Muskingum County (Fig. 17).

Environmental interpretation. The red color in shale is controlled by the oxidation state of iron (Tomlinson, 1916), is commonly diagenetic rather than detrital in origin (van Houten, 1973), and typically forms locally rather than from red source beds (Berner, 1971). A general absence of organic matter also

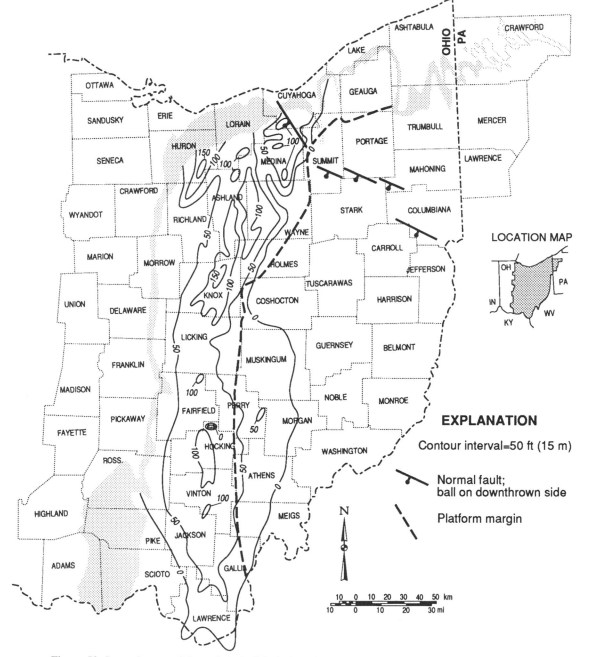

Figure 53. Isopach map of the red shale lithofacies of the Bedford Formation in Ohio and northwestern Pennsylvania (after Pepper et al., 1954).

favors preservation of red color (Potter et al., 1980). Although red shale is commonly a terrestrial deposit (e.g., McBride, 1974), marine examples also are common (e.g., Ziegler and McKerrow, 1975). Marine fossils and bioturbation are absent in the red shale lithofacies, as are exposure indicators such as mudcracks and root structures. However, the siltstone beds are identical to those in the gray shale lithofacies. Therefore, the red shale lithofacies was probably deposited not in terrestrial environments, as suggested by Pepper et al. (1954), but in marine environments, as suggested by Lewis (1988).

Siltstone beds help constrain the origin of the red shale because the associated gray shale caps apparently represent mud deposited during the waning phases of storms. Hence, the red shale is interpreted to be a product of hemipelagic background sedimentation. Preservation of gray shale caps further suggests that the red color is predepositional. Some relationships, however, support the diagenetic reduction of red mud to gray mud. For example, gray shale halos around organic grains apparently represent reduction of iron by decomposing organic matter. Additionally, the gray shale at the top of the lithofacies, which follows the base of the Berea regardless of deformation and erosion structures, may be related to reducing pore water that formed marcasite cement at the base of the Berea (cf. Berner, 1969; Maynard and Lauffenburger, 1978).

A key problem in interpreting the red shale lithofacies is identifying the source of the mud. The elongate geometry of the red shale (Fig. 53) indicates that it prograded southward in advance of the Cussewago delta front (i.e., quarrystone and deformed silty-sandstone lithofacies) (Fig. 17). Therefore, the shale is in a sense a distal prodelta deposit. However, the onlapping relationship of the red shale to the Euclid Member and the Cussewago–Second Berea Sandstone (Figs. 7, 10) indicates that the shale is mainly a transgressive facies. The soft, plastic character of the shale, moreover, suggests that fluid red mud helped lay the foundation for diapirism in northern Ohio by withdrawal from below subsiding quarrystone sand masses.

Potassium-argon dates from detrital mica in Bedford siltstone supports ultimate provenance of a large part of the Bedford-Berea sequence in the Acadian orogen (Aronson and Lewis, 1992). The red mud was probably not derived from the widespread Catskill redbeds in the east, however, because redbeds are absent in the proximal shelf and fluvial-deltaic deposits of the Bedford-Berea, which locally incise Catskill redbeds (Pepper et al., 1954). Redbeds are scarce, if present anywhere, in the Bedford-Berea of the Michigan basin (Newcombe, 1933; Gutschick and Sandberg, 1991). However, the position of thick red shale near the platform margin in Ohio and the overall geometry of the deposit indicate that mud was transported southward by currents that hugged the eastern edge of the western basin. Perhaps the best interpretation is that the shale had diverse sources north of the Appalachian basin and that regional onlap facilitated erosion and winnowing of preexisting Acadian redbeds in the source area.

Black shale lithofacies

Characteristics. The black shale lithofacies includes the Cleveland Shale and the Sunbury Shale (Figs. 1, 5). The Cleveland Shale is present throughout the western outcrop belt and is approximately 12–19 m (40–60 ft) thick (Fig. 7). Along the northern outcrop belt, the Cleveland thins eastward from approximately 19 m (60 ft) in Lorain and Huron Counties to less than 7 m (20 ft) in Trumbull County. However, black shale thinner than 6 m (18 ft) was not observed in outcrop in northeastern Ohio, and rapid thinning of the Bedford Formation in this area suggests that the Cleveland may be truncated by the Cussewago Sandstone.

Black shale is fissile and lithologically homogeneous (Fig. 53D), save for some thin gray shale, siltstone, and cone-in-cone limestone beds, and for marcasite nodules. Although the shale generally is thinly laminated and lacks bioturbation, burrows commonly penetrate black shale beds from overlying gray shale beds. However, some black Cleveland Shale in northern Ohio is bioturbated to the extent that laminae are not visible (Jordan, 1984). Few macrofossils are common in the black shale, although the Cleveland is famous for placoderm lagerstätten (e.g., Hyde, 1926). In-situ faunas containing brachiopods and bivalves also are present locally in the shale and have been discussed by Hlavin (1976) and Pashin and Ettensohn (1992a).

The black Cleveland Shale is present throughout the western basin and follows the platform margin closely in northern Ohio (Fig. 54). In east-central and southeastern Ohio, the Cleveland extends onto the eastern platform as far east as Tuscarawas County and as far south as Gallia County. In the western basin, the shale ranges in thickness from less than 3 to more than 30 m (10–100 ft). Here, three black shale depocenters are marked by the 24-m (80-ft) contour. Each depocenter coincides with Bedford-Berea sequence thicker than 38 m (125 ft) and with Bedford Shale thicker than 30 m (100 ft), although depocenters are offset slightly (Figs. 11, 54). Definition of the eastern limit of the Cleveland Shale in northeastern Ohio is difficult, but the isopach pattern (Fig. 54) suggests that the shale is truncated by the Cussewago Sandstone.

The Cleveland thins sharply at the platform margin and is generally thinner than 6 m (20 ft) on the platform (Figs. 9, 10, 54). Eastward extension of the black shale onto the platform in Tuscarawas County coincides with an embayment in the western margin of the Cussewago–Second Berea Sandstone (Fig. 12) and is approximately 32 km (20 mi) north of an eastward deflection of the pinchout of the red shale (Fig. 54). In Muskingum County, the Cleveland crosses the Cussewago–Second Berea siltstone belt, and from Morgan County southward, the pinchout follows the eastern margin of the belt.

The Sunbury Shale was not examined in detail in this investigation, but the shale is present throughout eastern Ohio. According to Prosser (1901, 1912), the Sunbury is thinner than 27 m (40 ft) throughout most of the western outcrop belt, is

Figure 54. Isopach map of the Cleveland Member of the Ohio Shale in Ohio and northwestern Pennsylvania.

only a few feet thick along the northern outcrop belt, and does not extend into Pennsylvania. The isopach map of Pepper et al. (1954) shows that the shale is thicker than 13 m (40 ft) in central Ohio. Throughout the remainder of the state, the shale is generally 7–13 m (20–40 ft) thick. The Sunbury has not been traced from Ohio into Pennsylvania, and the eastern pinchout of the shale approximately follows the Ohio state line from Ashtabula County to Jefferson County.

Environmental interpretation. Rich (1951a,b) recognized the importance of determining the depositional environment of

the Devonian-Mississippian black shale, because black fissile shale forms areally extensive marker units and provides a key for correlating North American basins (e.g., Fig. 5). He also pointed out that, because black shale units are so widespread and lithologically distinctive, they commonly serve as frames of reference about which regional stratigraphy is developed, and cited the Appalachian basin as an example. Contrary to many early studies, Rich argued in favor of a deep-water origin for the black shale, recognizing that coarser clastic intervals such as the Bedford-Berea sequence prograded into the black shale environment. Modern researchers (e.g., Byers, 1977; Ettensohn, 1985b; Kepferle, 1993) continue to propose a deep-water origin, citing modern analogs from oxygen-deficient basin floors of the Black Sea, the Sea of Azov, and the basins of the California Borderland (Caspers, 1957; Hülsemann and Emery, 1961).

The black fissile shale of the Cleveland and Sunbury is typical of deposits formed on basin floors within the anaerobic zone (Ettensohn and Elam, 1985). However, rare in-situ fossil assemblages (Hlavin, 1976) and locally abundant bioturbation (Jordan, 1984) in northern Ohio indicate that part of the basin floor was at times oxygenated enough to support some benthos. On the basis of directional data and gradation into underlying Olmsted shale and siltstone, Lewis (1988) suggested that some Cleveland Shale was not deposited far below storm wave base. Additionally, the Bedford sharply overlies the Cleveland throughout Ohio, and intertonguing between the gray and black shale is restricted to Kentucky, indicating that regional oxygenation of the basin floor was rapid, perhaps even abrupt, in Ohio.

The Cleveland Shale has generally been characterized as a transgressive, pelagic deposit on the basis of an onlapping relationship to the Catskill wedge (Fig. 5), fine grain size, fissility, and high kerogen content. However, Lewis and Schweitering (1971) identified the three black shale depocenters (Fig. 54) and suggested that input from terrigenous sources influenced the distribution of the Cleveland. Bedford-Berea depocenters are slightly offset from Cleveland depocenters (Figs. 11, 54), but thickening of the Bedford-Berea sequence above thick Cleveland Shale in the western basin provides evidence for control of sedimentation by differential subsidence and compaction.

Cross section F-F′ (Fig. 10) indicates that topographic relief along the platform margin was limited. Although sediment thickness increases westward by approximately 30 m (100 ft), all facies extend from the basin onto the platform. Hence, thickening of the Cleveland Shale and the Bedford-Berea sequence at the platform-basin transition was as strongly influenced by differential subsidence as by preexisting topography along the Chagrin shelf-to-basin transition. Restriction of thick Cleveland Shale to the western basin, however, indicates that the platform margin had enough relief to constrain the position of the pycnocline, at least for a time, thereby enabling accumulation of a large volume of compactible, organic-rich mud during Cleveland-Chagrin deposition. Therefore, differential compaction of black organic-rich mud and gray organic-poor mud may have acted in concert with reactivation of the Grenville suture in central Ohio to accentuate formation of the western basin.

Extension of black shale eastward into Coshocton County evidently represents encroachment of anoxic, basin-floor environments onto the eastern platform at the close of Cleveland-Chagrin deposition. Whereas the platform-basin transition may have been influenced in a major way by differential compaction and depositional topography inherited from Chagrin sedimentation, the eastern limit of the Cleveland may represent more subtle compactional and topographic control, because it is south of a major mud-rich lobe of the Chagrin (Fig. 54). Hence, eastward extension of the Cleveland is interpreted to represent oxygen-deficient, basin-floor sedimentation in the low area between older Chagrin sediment lobes.

DISCUSSION

Depositional history and paleogeography

Having identified the environments of deposition of each lithofacies, Bedford-Berea depositional history can be determined, and stratigraphic and sedimentologic data can be synthesized into a regional paleogeographic model. In the following discussion, a two-phase paleogeographic model is presented for the Bedford-Berea sequence (Figs. 55, 56). The principal constructive and destructive elements of the Bedford-Berea sequence were identified and integrated into a pair of reconstructions representing episodes of basin filling (time 1) and delta destruction (time 2).

Basin filling (time 1). The onlapping, intertonguing relationship of the Cleveland Shale to the Catskill clastic wedge is interpreted to represent regional transgression and slow destruction of the Catskill delta complex (Dennison, 1985). By contrast, the sharp, downlapping relationship of the Bedford-Berea sequence to the Catskill wedge (Fig. 5) indicates that the Bedford-Berea represents a major seaward shift of coastal onlap that facilitated widespread exposure and erosion of Catskill sediment. Exposure apparently led to incision of valleys by the Cussewago, Gay-Fink, and Cabin Creek fluvial systems and formation of an erosional unconformity on the eastern platform. Consequently, erosion of older clastic sediment apparently provided a large proportion of the sediment that was deposited in the western basin, and early Bedford-Berea deposition can be characterized as a major episode of delta construction and basin filling (Fig. 57).

The Bedford-Berea has the geometry of a lowstand wedge and is thus the product of a major forced regression (Posamentier and Vail, 1988; Jervey, 1992; Posamentier et al., 1992). The forcing mechanism for Bedford-Berea lowstand is a matter for speculation, but recent evidence points toward glacial eustasy. Although the Appalachian basin was in the arid southern tradewind belt at an approximate latitude of 25°S during the Late Devonian (Scotese, 1990), high-latitude continental glacial advances were taking place in Gondwana (Caputo and Crowell, 1985). Palynologic evidence from Bolivia suggests that one of those advances occurred just prior to the end of the Devonian (Vavrdová et al., 1991) and thus may have been coeval with Bedford-Berea lowstand. Regardless of the mechanism, the latest Devonian regression appears to have been extremely widespread, and lowstand deposits that appear to be equivalent to the Bedford-Berea have been identified in the Dinant synclinorium of Belgium (van Steenwinkel, 1990, 1993).

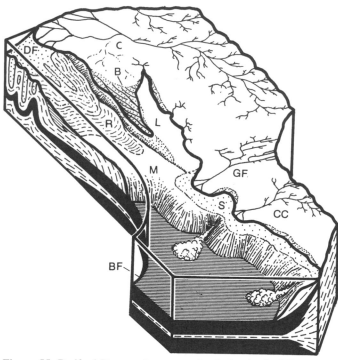

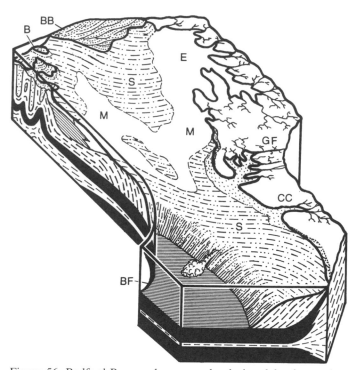

Figure 55. Bedford-Berea paleogeography during basin filling (time 1). Basin filling was characterized by fluvial systems that eroded the Catskill wedge and supplied prograding deltaic and shelf sediment to the western basin. C = Cussewago fluvial-deltaic system; DF = Cussewago delta front with diapiric structures; B = delta-destructive beach-barrier system; L = Lagoon; R = marine red mud; GF = Gay-Fink fluvial-deltaic system; CC = Cabin Creek fluvial-deltaic system; S = sandy and silty shelf; M = muddy shelf; BF = basin floor.

Figure 56. Bedford-Berea paleogeography during delta destruction (time 2). At this time, delta-front deposits in the western basin were uplifted and reworked, and a shelf silt blanket prograded back toward the incised valleys on the rapidly subsiding eastern platform where estuaries were forming. B = delta-destructive beach systems; BB = intertidal bar belt; E = estuarine embayment; GF = Gay-Fink estuary; CC = Cabin Creek estuary; S = sandy and silty shelf; M = muddy shelf; BF = basin floor.

Stratigraphic relationships indicate that among the earliest events in the basin-filling episode were oxygenation of the basin floor and establishment of the Euclid shelf (Fig. 57). The sharp Cleveland-Bedford contact in Ohio indicates that oxygenation of the basin floor was fairly rapid. Paleocurrents in the Bedford Shale of northern Ohio (gray shale lithofacies), which includes the Euclid and Sagamore Members (siltstone lithofacies) support a southwestward slope (Fig. 16), suggesting that the northeastern part of the western basin was troughlike and was filled from the east and northeast.

Most subsequent events in the northern part of the Appalachian basin can be related to development of the Cussewago fluvial-deltaic system (Fig. 55), but determination of the precise sequence of events is difficult. For example, incision of the Cussewago-Second Berea fluvial system into the Chagrin Shale and possibly the Cleveland Shale suggests that the early sedimentary record of the northern platform has been erased. Even so, the relationship of the red shale lithofacies to adjacent strata affords some constraint and further provides evidence for two major episodes of coastal onlap as the western basin filled with sediment.

Development of Cussewago–Second Berea beach and shelf deposits along the platform margin in southeastern Ohio

probably was early because part of the belt rests on the Cleveland Shale and is overlain by red shale (Figs. 8, 11, 12). The onlapping relationship of the red shale to the Euclid Member and the Cussewago–Second Berea siltstone belt indicates that the shelf aggraded or even retrograded (Fig. 57). Hence, for a time, sediment influx from the Cussewago fluvial system in the southeast waned and gave way to a major influx of mud (red shale lithofacies) that prograded from the north and filled the western basin. Perhaps an increasing rate of onlap following formation of the Euclid shelf caused a temporary waning of coarse sediment influx and facilitated reworking and winnowing of sediment north of the Appalachian basin that may have provided a source for the red mud. Following red shale deposition the Cussewago delta was apparently rejuvenated, and the second episode of coastal onlap commenced. During this episode, the Cussewago delta front (quarrystone and deformed silty-sandstone lithofacies) prograded onto the unstable red mud, thereby forming the diapiric structures of north-central Ohio (Figs. 55, 57).

In the Michigan basin, the Bedford Shale and Berea Sandstone were deposited as the progradational Thumb delta of Cohee (1965) and thus appear to be linked to the basin-filling episode in the Appalachian basin. The Thumb delta is generally thought to have prograded southwest from Canada (Pepper et al., 1954; Potter and Pryor, 1961), but Gutschick and Sandberg (1991) identified some possible northwest-prograding elements. Local evidence for northwest progradation suggests that the Cussewago delta in some manner may have delivered sediment into the Michigan basin through the Chatham sag, which separates the Findlay and Algonquin arches. If a connection indeed existed between the Thumb and Cussewago deltas, then a considerable portion of the Bedford-Berea sequence in the Michigan basin must have been deposited during the second episode of coastal onlap.

Whereas the Cussewago delta dominated events in the northern part of the Appalachian basin, most events in the southern part were related to the Gay-Fink and Cabin Creek trends (Fig. 55). The basin-filling episode was probably characterized by a constructive phase in this area although, as already stated, the primary evidence for construction is extension of trunk channels beyond the main trends. The trends are depicted as terminating in shoal-water, platform-margin deltas that supplied the mud and silt of the Bedford Shale and Berea Siltstone to the western basin of Kentucky.

Progradation of the Bedford Shale and Berea Siltstone in Kentucky (Figs. 20, 49) apparently was synchronous with filling of the northern part of the western basin, as indicated by intertonguing of siltstone and red shale in southernmost Ohio and northeastern Kentucky. Whereas oxygenation of the basin floor in Ohio was apparently abrupt, intertonguing of black and gray shale in Kentucky indicates that oxygenation was more

gradual and was characterized by progradation of mud-rich storm deposits and turbidite aprons of the gray shale lithofacies (Pashin and Ettensohn, 1987, 1992a,b). Siltstone deposition apparently commenced with deposition of structurally influenced turbidite systems and culminated in construction of a prograding storm-dominated shelf (Pashin and Ettensohn, 1987) (Fig. 55).

Delta destruction (time 2). Numerous events occurred during time 2, including delta destruction, shoaling, estuary formation and, ultimately, regional transgression (Fig. 56). The principal events in the northern part of the Appalachian basin evidently were destruction of the Cussewago delta, opening of the estuarine embayment, and development of the Berea shelf silt blanket. In the southern part of the basin, progradation of the Gay-Fink and Cabin Creek deltas and the Berea Siltstone eventually gave way to formation of estuaries and development of an open shelf that was deficient of sediment. Similar events were recognized by Pepper et al. (1954).

Delta destruction and estuary formation in northeastern Ohio and northwestern Pennsylvania included reworking of fluvial and deltaic sediment and deposition of beach and intertidal bar deposits of the solitary-crossbed and low-angle crossbed lithofacies. One consequence of estuary formation was development of a new subbasin in which the Knapp Conglomerate and the Corry Sandstone accumulated. Delta destruction in north-central Ohio included reworking of the Cussewago delta front and deposition of the beach deposits of the low-angle crossbed lithofacies around inactive delta-front sand bodies and

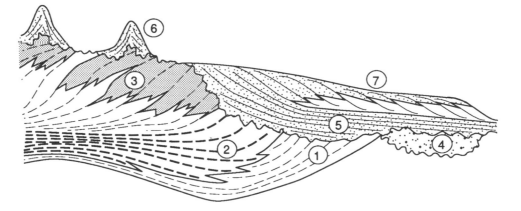

BASIN FILLING
1. Establisment and aggradation of Euclid shelf
2. Deposition of red mud on lower slope
3. Progradation of Cussewago delta; diapirism
4. Erosion of platform by Cussewago fluvial system

DELTA DESTRUCTION
4. Alluviation of Cussewago fluvial system
5. Reworking of fluvial-deltaic sediment,
   formation of estuarine embayment
6. Reworking of Cussewago delta front, beach formation
7. Progradation of siltstone blanket into estuarine embayment

Figure 57. Stratal geometry of the Bedford-Berea sequence related to basin-filling and delta-destruction episodes.

deformed shale masses (quarrystone and deformed silty-sand-stone lithofacies) (Figs. 56, 57). Reworking of the Cussewago delta apparently caused southward and eastward progradation of the shelf-silt blanket in northern Ohio (siltstone lithofacies) and development of a regionally extensive storm-dominated shelf. Shelf development culminated in deposition of the wide-spread silty sandstone bed of the low-angle crossbed litho-facies, and paleocurrent data in the gray shale lithofacies and the siltstone lithofacies (Figs. 16, 45) indicate that the shelf had a uniform south-southwest paleoslope.

On the Mississippi Delta, small delta-destructive beaches form by reworking the margins of delta lobes as they are aban-doned and subside below sea level (Coleman and Gagliano, 1964; Penland and Boyd, 1985). Destruction of the Cussewago delta was unusual, because facies relationships indicate that a large part of the delta front was reworked and that a large quan-tity of derived sediment was transported a considerable distance (Fig. 60). Whereas exposure and reworking took place in the northern part of the western basin, transgression and estuary formation occurred on adjacent parts of the eastern platform. Hence, paleoslope apparently reversed after delta construction and basin filling (time 1) and, although some sediment may have continued to enter the Appalachian and Michigan basins from the northeast, sediment derived from the Cussewago fluvial-deltaic system was reworked and transported back toward the source (Fig. 57). These events suggest that differen-tial uplift of the of the western basin and subsidence of the east-ern platform occurred in the northern part of the study area near the close of Bedford-Berea deposition.

Stratigraphic relationships along the northern outcrop demonstrate that delta destruction and embayment formation were diachronous (Fig. 57). Shore-zone deposits of the solitary-crossbed and low-angle crossbed lithofacies rest discon-formably on fluvial and deltaic sediment, including the red shale, reflecting the early stages of delta destruction and trans-gression. Near the close of Bedford-Berea deposition, the sand of the solitary-crossbed lithofacies was overlapped by mud in response to subsidence of the embayment. Continued uplift, reworking of the delta front, and beach formation (low-angle crossbed lithofacies) in the west apparently provided the blan-ket of silty sand (upper part of solitary-crossbed lithofacies) and silt (Berea Siltstone) that prograded eastward over the mud into the estuarine embayment (Figs. 56, 57).

Although delta destruction and estuary formation predomi-nated in the northern part of the Appalachian basin, delta con-struction and progradation of the Berea Siltstone may have continued for a time in the south. After the Berea Siltstone reached its maximum extent in the western basin, the coastline in West Virginia was inundated, and the estuarine Gay-Fink and Cabin Creek trends were preserved by transgression (Fig. 56). Following these events, transgression apparently hastened, the shelf no longer received sediment, and the burrowed, pyritic con-densation layer that forms top of the Bedford-Berea sequence throughout most of the study area accumulated. As transgression

continued, the Bedford-Berea cycle of deposition of was com-plete, and anoxic black shale deposition (Sunbury Shale) again prevailed throughout much of the Appalachian basin, thus fore-shadowing progradation of the Pocono clastic wedge.

Tectonic implications

Analysis of depositional history and paleogeography indi-cates that the Bedford-Berea lowstand wedge is in large part the product of relative sea-level variation. Among the most impor-tant effects related to sea level are widespread erosion of the Catskill clastic wedge and preservation of progradational deltaic sequences in the western basin and aggradational valley-fill sequences on the eastern platform. However, evi-dence for widespread structural control of sedimentation, such as incision of structurally controlled paleovalleys and the unusual nature of delta destruction in northern Ohio, suggests that tectonism was also a major control on Bedford-Berea deposition.

Acadian tectonism migrated southwest in response to sequential collision with the St. Lawrence, New York, and Vir-ginia continental promontories. Consequently, the locus of deformational loading and sedimentation progressively shifted southward, and the geometry of the Acadian foreland basins changed in response to these large-scale tectonic movements (Rodgers, 1967; Thomas, 1977; Ettensohn, 1985a, 1987; Faill, 1985). Comparing Catskill and Bedford-Berea paleocurrents helps confirm changes in foreland basin geometry related to Acadian tectonism (Fig. 58). Devonian turbidites of the Catskill clastic wedge were transported due west (McIver, 1970; Lun-degard et al., 1985; Lewis, 1988), whereas open-marine sedi-ment of the Bedford-Berea sequence was generally deposited on a southwestward paleoslope.

Change from a westward to a southwestward paleoslope suggests tilting of the Appalachian basin. In addition to transfer of loading strain from the New York promontory to the Virginia promontory, prolonged postorogenic uplift in the northern Appalachian orogen related to cessation of compressional tec-tonics and formation of pull-apart basins (Bradley, 1982) may have contributed to tilting of the Appalachian basin (Fig. 6). Introduction of siliciclastic sediment into the Appalachian and Michigan basins from the northeast apparently also occurred in response to uplift in the northern part of the Acadian orogen (de Witt and McGrew, 1979) and may have been another manifes-tation of regional tilting.

Foreland basins and their associated arch regions, which differentiate these basins from cratonic basins, form in response to the emplacement of thrust loads in the adjacent orogen (Beaumont, 1981; Jordan, 1981; Quinlan and Beaumont, 1984; Beaumont et al., 1988). During active deformational loading, which may occur in less than 1 m.y., the peripheral bulge migrates away from the orogen, forming a regional unconfor-mity. As the bulge migrates away from the orogen and the lithosphere begins to accommodate the tectonic load, the flex-

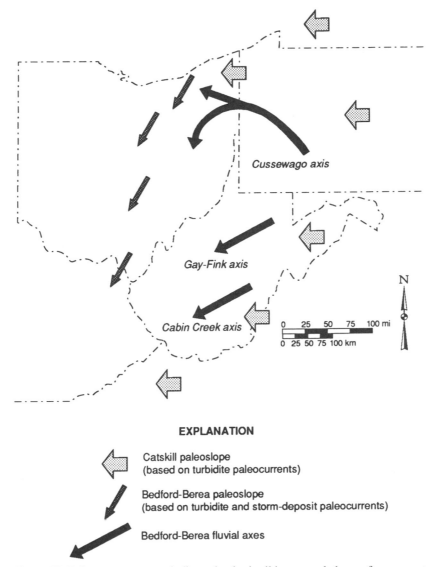

Figure 58. Paleocurrent vectors indicate that basin tilting caused change from a westward paleoslope during Catskill deposition to a southwestward paleoslope during Bedford-Berea deposition. This change is interpreted to reflect southward migration of the locus of Acadian deformational loading from the New York promontory to the Virginia promontory in concert with prolonged relaxational uplift in the northern Appalachians.

ural moat subsides faster than sediment is supplied from the emerging fold-and-thrust belt, thereby forming a thin, transgressive sequence.

According to the temperature-dependent viscoelastic models of Quinlan and Beaumont (1984) and Beaumont et al. (1988), isostatic adjustment of the crust to deformational loading may continue for tens of millions of years and is termed relaxation. As the relaxing crust accommodates a thrust load, the flexural moat continues to subside. However, the flexural moat narrows as the peripheral bulge lifts upward and migrates back toward the orogen (Fig. 59). During this phase, recently emplaced thrust loads are eroded and the foreland basin is filled with sediment, thus resulting in major depositional sequences

such as the Catskill and Pocono clastic wedges (Fig. 5). At the same time, erosion and reworking may take place on the uplifted bulge.

Tectonism that is expressed elastically or viscoelastically in the upper mantle may be expressed as brittle deformation near the surface, particularly as normal faults formed by flexural extension (Bradley and Kidd, 1991). Reactivation of preexisting basement faults may further localize stress, thereby influencing basin subsidence and bulge migration (Waschbusch and Royden, 1992). Therefore, understanding the structural framework is especially important for accurate modeling of flexural influences on facies architecture and paleogeography in foreland basins. Evidence for reactivation of Grenvillian and Iapetan

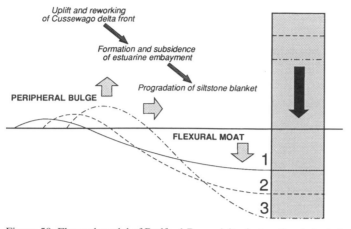

Figure 59. Flexural model of Bedford-Berea delta destruction (adapted from Quinlan and Beaumont, 1984). As the peripheral bulge was uplifted, distal deltaic sediment was reworked and deposited as a shelf siltstone blanket that prograded into the subsiding flexural moat where the estuarine embayment was forming.

basement structures during Bedford-Berea deposition suggests that reactivation of basement structures was a critical component of flexural tectonism in the Appalachian foreland basin.

A minor episode of bulge moveout associated with deformational loading in the Virginia promontory may have contributed to development of the unconformity below the incised valley fills on the eastern platform. However, that unconformity is restricted to the platform and is much more localized than the major regional unconformities that are commonly associated with bulge moveout (Quinlan and Beaumont, 1984). Therefore, eustasy was a more important forcing mechanism than tectonism for developing the unconformity and associated lowstand wedge. Although unconformity development and basin filling can be explained by eustasy, a tectonic mechanism is required to explain the unusual nature of Bedford-Berea delta destruction because deltaic sediment was exposed and reworked as the valley axis that fed the delta was inundated.

Indeed, Bedford-Berea delta destruction appears to record flexural modification of the lowstand wedge (Fig. 59). Reworking of the Cussewago delta may have been a response to uplift and migration of the peripheral bulge, whereas development of the estuarine embayment and progradation of the shelf-silt blanket may have been a response to subsidence of the flexural moat. During the earlier phase of basin filling, the thickness and distribution of facies were closely related to the platform margin. By the start of transgression and delta destruction, however, the western basin was effectively filled with sediment. Once the basin was full and water depth was shallow, topography and sediment distribution may have been sensitive to small vertical tectonic movements, perhaps of 10 m (33 ft) or less. Estuary formation in the Gay-Fink and Cabin Creek trends may also have been influenced by subsidence of the flexural moat, but uplift and migration of the peripheral bulge evidently did not affect shelf and slope sedimentation in Kentucky, perhaps

because the southern part of the western basin remained sufficiently below fair-weather wave base to be protected from reworking.

SUMMARY AND CONCLUSIONS

The Bedford-Berea sequence was deposited in the Appalachian basin during the Acadian orogeny and accumulated in response to a forced regression that followed deposition of the Catskill clastic wedge. Glacial-eustatic forcing of sea level apparently caused widespread exposure and erosion of the Catskill wedge, thus providing a significant source of siliciclastic sediment that was deposited in the distal parts of the Appalachian foreland basin. Lithofacies analysis indicates that the Bedford-Berea represents a spectrum of depositional systems, including fluvial, deltaic, estuarine, beach, storm-dominated shelf, turbidite, and oxygen-deficient basin-floor systems.

Among the salient features of Bedford-Berea deposition were an eastern platform, on which the sequence is generally thinner than 75 ft (23 m), and a western basin, in which the sequence is generally thicker than 125 ft (38 m). A large part of the eastern platform was characterized by erosion of Catskill sediment and subsequent deposition of aggradational, transgressive, valley-fill sequences. In contrast, the western basin was characterized largely by deposition of progradational, regressive, delta, and shelf sequences that conformably overlie the most distal part of the Catskill wedge.

Structural and sedimentologic evidence indicates that the platform margin has a multifold origin. Relict topography apparently played a major role in developing the platform and basin because the platform margin is developed where Catskill turbidites (Chagrin Shale) pass into basin-floor mud (Cleveland Shale). However, facies relationships demonstrate that topographic relief was limited along many parts of the platform margin and that segregation of the platform and basin was accentuated by differential subsidence. One cause of subsidence of the basin relative to the platform apparently was enhanced compaction of organic-rich Cleveland and Huron mud relative to organic-poor Chagrin mud and silt. In central Ohio, reactivation of a Grenville basement structure may have also contributed to differential subsidence.

Analysis of depositional history and paleogeography indicates that Bedford-Berea sedimentation can be divided into two major episodes: basin filling and delta destruction. Basin filling was characterized by formation of three structurally influenced fluvial-deltaic systems that are represented by the Cussewago–Second Berea Sandstone, the Gay-Fink trend, and the Cabin Creek trend. These systems apparently had sources near the Virginia promontory, eroded the Catskill wedge, and supplied distal deltaic and shelf sediment to the western basin. Additionally, a minor transgression may have caused reworking and winnowing of sediment north of the study area and deposition of red Bedford Shale. Following this event, regression apparently resumed, and the Cussewago delta prograded onto undercompacted red mud, thereby causing extensive diapirism in north-central Ohio.

Delta destruction was a time of shoaling, development of a regionally extensive storm-dominated shelf, and formation of estuaries. Although simple transgressive estuarine sequences were preserved in the Gay-Fink and Cabin Creek trends of West Virginia, delta destruction in northern Ohio and northwestern Pennsylvania was an unusual event that included subsidence of the estuarine embayment, uplift and reworking of part of the Cussewago delta, and progradation of a shelf silt blanket from the western basin onto the eastern platform. Paleoslope in the northern part of the Appalachian basin apparently reversed after the western basin was filled and, although some sediment may have continued to enter the basin from the north, sediment derived from the Cussewago–Second Berea fluvial deltaic system was evidently reworked and transported back toward the southeast.

Paleocurrent data indicate southward tilting of the basin that was probably related to a southward migrating locus of thrust loading during the Acadian orogeny coupled with prolonged relaxational uplift in the northern Appalachian orogen. Tectonic reorganization of the basin may have further facilitated widespread reactivation of basement faults. Flexural relaxation related to Acadian orogenesis may have contributed to the unusual nature of Bedford-Berea delta destruction. Reworking of the Cussewago delta can be explained in terms of uplift and migration of a peripheral bulge, whereas development of the estuarine embayment and progradation of the shelf-silt blanket can be explained by concomitant subsidence of a flexural moat.

The primary difference between this study and the classic studies of Rich (1951a,b) and Pepper et al. (1954) is the realization that the Bedford-Berea sequence was deposited in a tectonically dynamic foreland basin that was undergoing flexural reorganization concurrently with eustatic changes of sea level. Moreover, tectonism, relict topography, differential compaction, and changing sea level appear to have functioned collectively to determine the complex depositional history and paleogeography of the Bedford-Berea sequence. Tectonism, relict topography, and differential compaction evidently acted in unison to provide sediment sources and establish the geometry of the sedimentary basin as well as that of the basin fill. By contrast, eustatic sea-level variation helped determine the position and rate of change of base level and was thus a critical factor that contributed to erosion of the Catskill wedge, transport of sediment into the basin from northern sediment sources, preservation of valley fills west of the limit of Appalachian thrusting, and restriction of thick progradational sequences to the western basin.

Reevaluation of the Bedford-Berea sequence demonstrates that epeiric sea-floor topography was much more elaborate than has commonly been acknowledged. One reason for this elaborate topography is that sedimentation in many foreland and cratonic basins was influenced by basement structures with varied orientations that responded to changes within orogenic belts and thus reflects a polyphase crustal history. The limited topographic relief of epeiric sea floors apparently made sediment distribution extremely sensitive to tectonic evolution of the sedimentary basin, evolving depositional topography, differential compaction, and sea-level fluctuation. Hence, interplay of these factors may have resulted in myriad patterns of erosion and deposition in the geologic record and, like the Bedford-Berea sequence, each depositional sequence has intricacies that must be identified before a thorough knowledge of foreland basins can be achieved.

REFERENCES CITED

Aigner, T., and Reineck, H.-E., 1982, Proximity trends in modern storm sands from the Helgoland Bight (North Sea) and their implications for basin analysis: Senckenbergiana Maritima, v. 14, p. 183–215.

Allen, J. R. L., 1963, The classification of cross-stratified units with notes on their origin: Sedimentology, v. 2., p. 93–114.

Allen, J. R. L., 1982, Sedimentary structures: their character and physical basis: Developments in Sedimentology, v. 30, part II, 663 p.

Ammerman, M. L., and Keller, G. R., 1979, Delineation of Rome trough in eastern Kentucky by gravity and deep drilling data: American Association of Petroleum Geologists Bulletin, v. 63, p. 341–353.

Aronson, J. L., and Lewis, T. L., 1992, Detrital mica K/Ar ages for Devonian-Pennsylvanian strata of the north-central Appalachian basin: Dominance of the Acadian orogen as provenance: Geological Society of America Abstracts with Programs, v. 24, p. A237.

Asquith, D. O., 1970, Depositional topography and major marine environments, Late Cretaceous, Wyoming: American Association of Petroleum Geologists Bulletin, v. 54, p. 1184–1224.

Asseez, L. O., 1969, Paleogeography of Lower Mississippian rocks of Michigan basin: American Association of Petroleum Geologists Bulletin, v. 53, p. 127–135.

Beaumont, C., 1981, Foreland basins: Geophysical Journal of the Royal Astronomical Society, v. 65, p. 291–329.

Beaumont, C., Quinlan, G., and Hamilton, J., 1988, Orogeny and stratigraphy: Numerical models of the Paleozoic in the eastern interior of North America: Tectonics, v. 7, p. 389–416.

Beets, D. J., Roep, Th. B., and deJong, J., 1981, Sedimentary sequences of the sub-recent North Sea coast of the western Netherlands near Alkmaar, *in* Nio, S.-D., ed., Holocene sedimentation in the North Sea Basin: International Association of Sedimentologists Special Publication 5, p. 133–145.

Berg, T. M., and Edmunds, W. E., 1979, The Huntley Mountain Formation: Catskill-to-Burgoon transition in north-central Pennsylvania: Pennsylvania Topographic and Geologic Survey Information Circular 83, 80 p.

Berg, T. M., McInerney, M. K., Way, J. H., and MacLachlan, D. B., 1983, Stratigraphic correlation chart of Pennsylvania: Pennsylvania Geological Survey, 4th Series, General Geology Report 75.

Berner, R. A., 1969, Migration of iron and sulphur within anaerobic sediments during early diagenesis: American Journal of Science, v. 257, p. 19–42.

Berner, R. A., 1971, Principles of Chemical Sedimentology: New York, McGraw-Hill, 240 p.

Bigarella, J. J., 1965, Sand-ridge structures from Paraña coastal plain: Marine Geology, v. 3, p. 269–278.

Bjerstedt, T. W., 1986, Regional stratigraphy and sedimentology of the Rockwell Formation and Purslane Sandstone based on the new Sideling Hill road cut, Maryland: Southeastern Geology, v. 27, p. 69–94.

Bjerstedt, T. W., and Kammer, T. W., 1988, Genetic stratigraphy and depositional systems of the Upper Devonian–Lower Mississippian Price-Rockwell deltaic complex in the central Appalachians, U.S.A.: Sedimentary Geology, v. 54, p. 265–301.

Boersma, J. R., and Terwindt, J. H. J., 1981, Neap-spring tide sequence of intertidal shoal deposits in a mesotidal estuary: Sedimentology, v. 28, p. 151–170.

Boswell, R., 1988, Stratigraphic expression of basement fault zones in northern West Virginia: Geological Society of America Bulletin, v. 100, p. 1988–1998.

Bouma, A. H., 1962, Sedimentology of some flysch deposits: Amsterdam, Elsevier, 168 p.

Bourgeois, J., 1980, A transgressive shelf sequence exhibiting hummocky stratification: The Cape Sebastian Sandstone (Upper Cretaceous), southwestern Oregon: Journal of Sedimentary Petrology, v. 50, p. 681–702.

Bownocker, J. A., 1906, The occurrence and exploitation of petroleum and natural gas in Ohio: Report of the Geological Survey of Ohio, v. 8, Bulletin 1, 325 p.

Bradley, D. C., 1982, Subsidence in late Paleozoic basins in the northern Appalachians: Tectonics, v. 2, p. 107–123.

Bradley, D. C., and Kidd, W. S. F., 1991, Flexural extension of the upper continental crust in collisional foredeeps: Geological Society of America Bulletin, v. 103, p. 1416–1438.

Brenner, R. L., 1978, Sussex Sandstone of Wyoming—Example of Cretaceous offshore sedimentation: American Association of Petroleum Geologists Bulletin, v. 62, p. 181–200.

Briggs, C., Jr., 1838, Report of C. Briggs, Jr.: Columbus, First Annual Report of the Geological Survey of Ohio, p. 71–98.

Broadhead, R. F., Kepferle, R. C., and Potter, P. E., 1982, Stratigraphic and sedimentologic controls of gas in shale—Example from the Upper Devonian of northern Ohio: American Association of Petroleum Geologists Bulletin, v. 66, p. 10–27.

Bruce, C. H., 1973, Pressured shale and related sediment deformation: Mechanism for development of regional contemporaneous faults: American Association of Petroleum Geologists Bulletin, v. 57, p. 878–886.

Bucher, W. H., 1919, On ripples and related sedimentary surface forms and their paleogeographic interpretation: American Journal of Science, v. 47, p. 241–269.

Burrows, V. C., 1988, Subsurface stratigraphy and paleoenvironmental interpretation of the Mississippian Berea Sandstone and Bedford Formation of Medina County, Ohio [M.S. thesis]: Kent, Ohio, Kent State University, 121 p.

Byers, C. W., 1977, Biofacies patterns in euxinic basins: A general model: Society of Economic Paleontologists and Mineralogists Special Publication 25, p. 5–17.

Caputo, M. V., and Crowell, J. C., 1985, Migration of glacial centers across Gondwana during Paleozoic Era: Geological Society of America Bulletin, v. 96, p. 1020–1036.

Carll, J. F., 1890, Seventh report on the oil and gas fields of western Pennsylvania: 2nd Pennsylvania Geological Survey, v. 15, 356 p.

Carter, J. L., and Kammer, T. W., 1990, Late Devonian and Early Carboniferous brachiopods (Brachiopoda, Articulata) from the Price Formation of West Virginia and adjacent areas of Pennsylvania and Maryland: Annals of Carnegie Museum, v. 59, p. 77–103.

Caspers, H., 1957, Black Sea and Sea of Azov, in Hedgpeth, J. W., ed., Treatise on marine ecology and paleoecology: Geological Society of America Memoir 67, p. 801–890.

Caster, K. E., 1934, The stratigraphy and paleontology of northwestern Pennsylvania; pt I: Stratigraphy: Bulletins of American Paleontology, v. 21, p. 1–185.

Chaplin, J. R., 1980, Stratigraphy, trace fossil associations, and depositional environments in the Borden Formation (Mississippian); in Geological Society of Kentucky Annual Field Conference Guidebook: Lexington, Kentucky Geological Survey, 114 p.

Clifton, H. E., Hunter, R. E., and Phillips, R. L., 1971, Depositional structures and processes in the non-barred high-energy nearshore: Journal of Sedimentary Petrology, v. 41, p. 651–670.

Cohee, G. V., 1965, Geologic history of the Michigan basin: Journal of the Washington Academy of Science: v. 55, p. 211–223.

Cohee, G. V., and Underwood, L. B., 1944, Maps and sections of the Berea Sandstone in the Michigan basin: U.S. Geological Survey Oil and Gas Investigation, Preliminary Map 17.

Coleman, J. M., 1982, Deltas—Processes of deposition and models for exploration (2nd ed.): Boston, International Human Resources Development Corporation, 124 p.

Coleman, J. M., and Gagliano, S. M., 1964, Cyclic sedimentation in the Mississippi river deltaic plain: Gulf Coast Association of Geological Societies Transactions, v. 14, p. 67–80.

Coogan, A. H., Heimlich, R. A., Malcuit, R. A., Bork, K. B., and Lewis, T. L., 1981, Early Mississippian deltaic sedimentation in central and northeastern Ohio, in Roberts, T. G., ed., Stratigraphy, sedimentology (Geological Society of America Cincinnati 1981 Field Trip Guidebooks, v. I): Falls Church, Virginia, American Geological Institute, p. 113–152.

Cooper, J. R., 1943, Flow structures in the Berea Sandstone and Bedford Shale of central Ohio: Journal of Geology, v. 51, p. 190–203.

Cotter, E., 1985, Gravel-topped offshore bar sequences in the Lower Carboniferous of southern Ireland: Sedimentology, v. 32, p. 195–213.

Craft, J. H., and Bridge, J. S., 1987, Shallow-marine sedimentary processes in the Late Devonian Catskill Sea, New York State: Geological Society of America Bulletin, v. 98, p. 338–355.

Culotta, R. C., Pratt, T., and Oliver, J., 1990, A tale of two sutures: COCORP's deep seismic surveys of the Grenville province in the eastern U. S. mid-continent: Geology, v. 18, p. 646–649.

Cushing, H. P., Leverett, F, and van Horn, F. R., 1931, Geology and mineral resources of the Cleveland District, Ohio: U.S. Geological Survey Bulletin 818, 138 p.

Davis, R. A., Fox, W. T., Hayes, M. O., and Boothroyd, J. C., 1972, Comparison of ridge and runnel systems in tidal and non-tidal environments: Journal of Sedimentary Petrology, v. 42, p. 413–421.

Demarest, D. F., 1946, Map of the Berea and Murrysville sands of northeastern Ohio, western Pennsylvania, and northernmost West Virginia: U.S. Geological Survey Oil and Gas Investigation, Preliminary Map 49.

Dennison, J. M., 1985, Catskill delta shallow marine strata: Geological Society of America Special Paper 201, p. 91–106.

de Witt, W., Jr., 1946, The stratigraphic relationship of the Berea, Corry, and Cussewago sandstones in northeastern Ohio and northwestern Pennsylvania: U.S. Geological Survey Oil and Gas Investigation, Preliminary Map 21.

de Witt, W., Jr., 1951, Stratigraphy of the Berea Sandstone and associated rocks in northeastern Ohio and northwestern Pennsylvania: Geological Society of America Bulletin, v. 62, p. 1347–1370.

de Witt, W., Jr., and McGrew, L. W., 1979, Appalachian basin region, in Craig, L. C., and Connor, C. W., eds., Paleotectonic investigations of the Mississippian System in the United States: U.S. Geological Survey Professional Paper 1010-C, p. 13–48.

de Witt, W., Jr., Roen, J. B., and Wallace, L. G., 1993, Stratigraphy of Devonian black shales and associated rocks in the Appalachian basin: U.S. Geological Survey Bulletin 1909-B, 57 p.

Dickinson, W. R., and 8 others, 1983, Provenance of North American Phanerozoic sandstones in relation to tectonic setting: Geological Society of America Bulletin, v. 94, p. 222–235.

Dillman, S. B., 1980, Subsurface geology of the Upper Devonian–Lower Mississippian black-shale sequence in eastern Kentucky [M.S. thesis]: Lexington, University of Kentucky, 72 p.

Dolly, E. D., and Busch, D. A., 1972, Stratigraphic, structural, and geomorphic factors controlling oil accumulation in Upper Cambrian strata of Ohio: American Association of Petroleum Geologists Bulletin, v. 56, p. 2335–2368.

Donaldson, A. C., and Schumaker, R. C., 1981, Late Paleozoic molasse of central Appalachians, in Miall, A. D., ed., Sedimentation and tectonics in alluvial basins: Geological Association of Canada Special Publication 23, p. 100–123.

Dott, R. H., and Bourgeois, J., 1982, Hummocky stratification: Significance of variable bedding sequences: Geological Society of America Bulletin, v. 93, p. 663–680.

Duncan, P. W., and Wells, N. A., 1992, The Mississippian Berea Sandstone at Bedford in northern Ohio: Tidal cyclicity, syndepositional deformation, and major bounding surfaces: Northeastern Geology, v. 14, p. 15–28.

Elam, T. D., 1981, Stratigraphy and paleoenvironmental aspects of the Bedford-Berea sequence and the Sunbury Shale in eastern and south-central Kentucky [M.S. thesis]: Lexington, University of Kentucky, 155 p.

Elliott, T., 1975, Deltas, *in* Reading, H. G., ed., Sedimentary environments and facies: New York, Elsevier, p. 97–142.

Ells, G. G., 1979, Stratigraphic cross sections extending from Devonian Antrim Shale to Mississippian Sunbury Shale in the Michigan basin: Michigan Geological Survey, Report of Investigations 22, 186 p.

Ettensohn, F. R., 1985a, The Catskill delta complex and the Acadian Orogeny: A model: Geological Society of America Special Paper 201, p. 39–50.

Ettensohn, F. R., 1985b, Controls on the development of Catskill delta complex basin-facies: Geological Society of America Special Paper 201, p. 65–77.

Ettensohn, F. R., 1987, Rates of relative plate motion during the Acadian orogeny based on the spatial distribution of black shales: Journal of Geology, v. 95, p. 572–582.

Ettensohn, F. R., and Barron, L. S., 1981, Depositional model for the Devonian-Mississippian black shales of North America: A paleogeographic-paleoclimatic approach, *in* Roberts, T. G., ed., Economic geology, structure (Geological Society of America Cincinnati 1981 Field Trip Guidebooks, v. II): Falls Church, Virginia, American Geological Institute p. 344–357.

Ettensohn, F. R., and Dever, G. R., eds., 1979, Carboniferous geology from the Illinois basin to the Appalachian basin through eastern Ohio and Kentucky: Lexington, Kentucky, University of Kentucky, Falls Church, Virginia, American Geological Institute, Field Trip no. 4, Ninth International Congress of Carboniferous Stratigraphy and Geology Guidebook, 293 p.

Ettensohn, F. R., and Elam, T. D., 1985, Defining the nature and location of a Late Devonian–Early Mississippian pycnocline in eastern Kentucky: Geological Society of America Bulletin, v. 96, p. 1313–1321

Ettensohn, F. R., and 8 others, 1988, Characterization and implications of the Devonian-Mississippian black shale sequence, eastern and central Kentucky, U.S.A.: Pycnoclines, transgression, regression, and tectonism, *in* McMillan, N. J., Embry, A. F., and Glass, D. J., eds., Devonian of the world, Proceedings, 2nd International Symposium on the Devonian System: Canadian Society of Petroleum Geologists Memoir 14, v. 2, p. 323–345.

Evans, G., 1965, Intertidal flat sediments and their environments of deposition in the Wash: Quarterly Journal of the Geological Society of London, v. 121, p. 209–245.

Faill, R. T., 1985, The Acadian orogeny and the Catskill delta: Geological Society of America Special Paper 201, p. 15–37.

Fairbridge, R. W., 1980, The estuary: Its definition and geodynamic cycle, *in* Olausson, E., and Cato, I., eds., Chemistry and biogeochemistry of estuaries: New York, John Wiley, p. 1–36.

Farrell, S. G., and Eaton, S., 1987, Slump strain in the Tertiary of Cyprus and the Spanish Pyrenees: Definition of palaeoslopes and soft-sediment deformation, *in* Jones, M. E., and Preston, R. M. F., eds., Deformation of sediments and sedimentary rocks: Geological Society of London Special Publication 29, p. 181–196.

Ferm, J. C., 1979, Introduction to collected papers and guidebook, *in* Ferm, J. C., and Horne, J. C., eds., Carboniferous depositional environments in the Appalachian region: Columbia, University of South Carolina, Carolina Coal Group, p. 1–9.

Ferm, J. C., and Weisenfluh, G. A., 1981, Cored rocks of the southern Appalachian coal fields: Lexington, Department of Geology, University of Kentucky, 93 p.

Fisher, J. H., 1980, Stratigraphy of the Upper Devonian-Lower Mississippian of Michigan: U.S. Department of Energy Report FE 2346-80, 21 p.

Fisk, H. N., 1961, Bar finger sands of the Mississippi delta, *in* Peterson, J. A., and Osmond, J. C., eds., Geometry of sandstone bodies—A symposium: Tulsa, Oklahoma, American Association of Petroleum Geologists, p. 29–52.

Folk, R. L., 1968, Petrology of sedimentary rocks: Austin, Hemphill's Bookstore, 170 p.

Frazier, W. J., and Schwimmer, D. R., 1987, Regional stratigraphy of North America: New York, Plenum Press, 719 p.

Goldring, R., and Bridges, P., 1973, Sublittoral sheet sandstones: Journal of Sedimentary Petrology: v. 43, p. 736–747.

Gray, J. D., 1982, Subsurface structure mapping of eastern Ohio, *in* Gray, J. D., and 8 others, eds., An integrated study of the Devonian-age black shales of eastern Ohio: Washington, D.C., U.S. Department of Energy, DOE/ET/12131-1399, p. 3.1–3.13.

Greenwood, B., and Sherman, D. J., 1986, Hummocky cross-stratification in the surf zone: Flow parameters and bedding genesis: Sedimentology, v. 33, p. 33–45.

Gutschick, R. C., and Sandberg, C. A., 1991, Late Devonian history of Michigan basin: Geological Society of America Special Paper 256, p. 181–202.

Hale, L., 1941, Study of sediments and stratigraphy of Lower Mississippian in western Michigan: American Association of Petroleum Geologists Bulletin, v. 25, p. 713–723.

Hamblin, A. P., and Walker, R. G., 1979, Storm-dominated shallow marine deposits: The Fernie-Kootenay (Jurassic) transition, southern Rocky Mountains: Canadian Journal of Earth Sciences, v. 16, p. 1673–1690.

Harrell, J. A., Hatfield, C. B., and Gunn, G. R., 1991, Mississippian System of the Michigan basin: Stratigraphy, sedimentology, and economic geology: Geological Society of America Special Paper 256, p. 203–219.

Harris, L. D., 1975, Oil and gas data from the Lower Ordovician and Cambrian rocks of the Appalachian Basin: U.S. Geological Survey Miscellaneous Investigation I-197-D, scale 1:2,500,000.

Heckel, P. H., 1972, Recognition of ancient shallow marine environments: Society of Economic Paleontologists and Mineralogists Special Publication 16, p. 226–286.

Herrick, C. L., 1888, Geology of Licking County, O., Pt. IV: Dennison University Scientific Laboratories Bulletin, v. 4, p. 97–130.

Hildreth, S. P., 1836, Observations on the bituminous coal deposits of the valley of the Ohio, and the accompanying rock strata; With notices of the fossil organic remains and the relics of vegetable and animal bodies, illustrated by a geological map, by numerous drawings of plants and shells, and by views of the scenery: American Journal of Science, v. 29, p. 1–154.

Hine, A. C., 1979, Mechanisms of berm development and resulting growth along a barrier spit complex: Sedimentology, v. 20, p. 333–351.

Hlavin, W. J., 1976, Biostratigraphy of Late Devonian black shales on the cratonic margin of the Appalachian geosyncline [Ph.D. thesis]: Boston, Boston University, 211 p.

Howard, J. D., and Lohrengel, C. F., II, 1969, Large non-tectonic deformational structures from Upper Cretaceous rocks of Utah: Journal of Sedimentary Petrology, v. 39, p. 1032–1039.

Hülsemann, J., and Emery, K. O., 1961, Stratification in Recent sediments of Santa Barbara basin as controlled by organisms and water character: Journal of Geology, v. 69, p. 279–290.

Hunter, R. E., Clifton, H. E., and Phillips, R. L., 1979, Depositional processes, sedimentary structures, and predicted vertical sequences in barred, nearshore systems, southern Oregon coast: Journal of Sedimentary Petrology, v. 49, p. 711–726.

Hyde, J. E., 1911, The ripples of the Bedford and Berea formations of central and southern Ohio: Journal of Geology, v. 19, p. 257–269.

Hyde, J. E., 1926. Collecting fossil fishes from the Cleveland Shale: Natural History (American Museum of Natural History), v. 26, p. 497–504.

Hyde, J. E., 1953, Mississippian formations of central and southern Ohio, *in*

Marple, M. F., ed.: Ohio Geological Survey Bulletin 51, 335 p.

Jervey, M. T., 1992, Siliciclastic sequence development in foreland basins, with examples from the western Canada foreland basin: American Association of Petroleum Geologists Memoir 55, p. 47–80.

Jillson, W. R., 1919, The pay oil sands of eastern Kentucky: Mineral and Forest Resources of Kentucky, v. 1, p. 334–367.

Jordan, M. R., 1984, Sedimentological examination of the Upper Devonian shales of northeastern Ohio [M.S. thesis]: Cleveland, Ohio, Case Western Reserve University, 236 p.

Jordan, T. E., 1981, Thrust loads and foreland basin evolution, Cretaceous, western United States: American Association of Petroleum Geologists Bulletin, v. 65, p. 2506–2520.

Kammer, T. W., and Bjerstedt, T. W., 1986, Stratigraphic framework of the Price Formation in West Virginia: Southeastern Geology, v. 27, p. 13–33.

Kepferle, R. C., 1977, Stratigraphy, petrology, and depositional environment of the Kenwood Siltstone Member, Borden Formation (Mississippian), Kentucky and Indiana: U.S. Geological Survey Professional Paper 1007, 49 p.

Kepferle, R. C., 1978, Prodelta turbidite fan apron in Borden Formation (Mississippian), Kentucky and Indiana, in Stanley, D. J., and Kelling, G., eds., Sedimentation in submarine fans, canyons, and trenches: Stroudsburg, Pennsylvania, Dowden, Hutchinson, and Ross, p. 224–238.

Kepferle, R. C., 1993, A depositional model and basin analysis for the gas-bearing black shale (Devonian and Mississippian) in the Appalachian basin: U.S. Geological Survey Bulletin 1909-F, 23 p.

Kepferle, R. C., Potter, P. E., and Pryor, W. A., 1981, Stratigraphy of the Chattanooga Shale (Upper Devonian and Lower Mississippian) in vicinity of Big Stone Gap, Wise County, Virginia: U.S. Geological Survey Bulletin 1499, 20 p.

Kindle, E. M., 1917, Recent and fossil ripple mark: Canada Geological Survey Museum Bulletin 25, p. 1–56.

Klein, G. deV., 1967, Comparison of recent and ancient tidal flat and estuarine sediments, in Lauff, G. H., ed., Estuaries: American Association for the Advancement of Science Publication 83, p. 207–218.

Kohout, D. L., and Malcuit, R. J., 1969, Environmental analysis of the Bedford Formation and associated strata in the vicinity of Cleveland, Ohio: Compass of ΣΓΕ, v. 46, p. 192–206.

Kohsiek, L. H. M., and Terwindt, J. H. J., 1981, Characteristics of foreset and topset bedding in megaripples related to hydrodynamic conditions on an intertidal shoal, in Nio, S.-D., ed., Holocene sedimentation in the North Sea basin: International Association of Sedimentologists Special Publication 5, p. 27–37.

Krause, R. G. F., and Geijer, T. A. M., 1987, An improved method for calculating the standard deviation and variance of paleocurrent data: Journal of Sedimentary Petrology, v. 57, p. 779–780.

Kreisa, R. D., and Bambach, R. K., 1973, Environments of deposition of the Price Formation (Lower Mississippian) in its type area, southwestern Virginia: American Journal of Science, v. 273-A, p. 326–342.

Kreisa, R. D., and Moiola, R. J., 1986, Sigmoidal tidal bundles and other tide-generated sedimentary structures of the Curtis Formation, Utah: Geological Society of America Bulletin, v. 97, p. 381–387.

Krumbein, W. C., and Sloss, L. L., 1963, Stratigraphy and sedimentation (2nd ed.): San Francisco, W. H. Freeman, 660 p.

Lamborn, R. E., Austin, C. R., and Schaaf, D., 1938, Shales and surface clays of Ohio: Ohio Geological Survey Bulletin 82, 96 p.

Larese, R. E., 1974, Petrology and stratigraphy of the Berea Sandstone in the Cabin Creek and Gay-Fink trends, West Virginia [Ph.D. thesis]: Morgantown, West Virginia University, 245 p.

Laury, R. L., 1971, Stream bank failure and rotational slumping: Preservation and significance in the geologic record: Geological Society Bulletin, v. 82, p. 1251–1266.

Leckie, D. A., and Walker, R. G., 1982, Storm- and tide-dominated shorelines in Cretaceous Moosebar–Lower Gates interval—Outcrop equivalents of deep basin gas trap in western Canada: American Association of Petro-

leum Geologists Bulletin, v. 66, p. 138–157.

Lené, G. W., and Owen, D. E., 1969, Grain orientation in a Berea Sandstone channel at South Amherst, Ohio: Journal of Sedimentary Petrology, v. 39, p. 737–743.

Lewis, T. L., 1968, Paleocurrent analysis of the Chagrin, Cleveland, Bedford, and Berea Formations of northern Ohio [abs.]: Geological Society of America Special Paper 115, p. 130–131.

Lewis, T. L., 1976, Late Devonian and Early Mississippian paleoenvironments, northern Ohio [abs.]: Geological Society of America Abstracts with Programs, v. 8, p. 220.

Lewis, T. L., 1988, Late Devonian and Early Mississippian distal basin-margin sedimentation of northern Ohio: Ohio Journal of Science, v. 88, p. 23–39.

Lewis, T. L., and Schweitering, J. F., 1971, Distribution of the Cleveland black shale in Ohio: Geological Society of America Bulletin, v. 82, p. 3477–3482.

Lilienthal, R. T., 1978, Stratigraphic cross-sections of the Michigan basin: Michigan Geological Survey, Report of Investigations 19, 36 p.

Lucius, J. E., and von Froese, R. R. B., 1988, Aeromagnetic and gravity anomaly constraints on the crustal geology of Ohio: Geological Society of America Bulletin, v. 100, p. 104–116.

Lundegard, P. D., Samuels, N. D., and Pryor, W. A., 1985, Upper Devonian turbidite sequence, central and southern Appalachian basin: Contrasts with submarine fan deposits: Geological Society of America Special Paper 201, p. 107–121.

Matthews, R. D., 1993, Review and revision of the Devonian-Mississippian stratigraphy in the Michigan basin: U.S. Geological Survey Bulletin 1909-D, 85 p.

Maynard, J. B., and Lauffenberger, S. K., 1978, A marcasite layer in prodelta turbidites of the Borden Formation (Mississippian) in eastern Kentucky: Southeastern Geology, v. 20, p. 47–58.

McBride, E. F., 1974, Significance of red, green, purple, olive, brown, and gray beds of Difunta Group, northeastern Mexico: Journal of Sedimentary Petrology, v. 44, p. 760–773.

McCave, I. N., and Geiser, A. C., 1979, Megaripples, ridges, and runnels in intertidal flats of the Wash, England: Sedimentology, v. 26, p. 353–369.

McCrory, V. L. C., and Walker, R. G., 1986, A storm and tidally-influenced prograding shoreline—Upper Cretaceous Milk River Formation of southern Alberta, Canada: Sedimentology, v. 33, p. 47–60.

McGregor, D. J., 1954, Stratigraphic analysis of Upper Devonian and Mississippian rocks in the Michigan basin: American Association of Petroleum Geologists Bulletin, v. 38, p. 2324–2356.

McIver, N. L., 1970, Appalachian turbidites, in Fisher, G. W., Pettijohn, F. J., Reed, J. C., Jr., and Weaver, K. N., eds., Studies of Appalachian geology—Central and southern: New York, Interscience, p. 69–81.

McKee, E. D., and Goldberg, M., 1969, Experiments on formation of contorted structures in mud: Geological Society of America Bulletin, v. 80, p. 231–244.

Mitchum, R. M., Jr., Vail, P. E., and Thompson, S., III, 1977, The depositional sequence as a basic unit for stratigraphic analysis: American Association of Petroleum Geologists Memoir 26, p. 53–62.

Moore, B. R., and Clark, M. K, 1970, The significance of a turbidite sequence in the Borden Formation (Mississippian) of eastern Kentucky and southern Ohio: Geological Association of Canada Special Paper 7, p, 211–218.

Morgan, J. P., 1961, Genesis and paleontology of the Mississippi River mudlumps: Louisiana Geological Survey Bulletin 35, 116 p.

Morgan, J. P., Coleman, J. M., and Gagliano, S. M., 1968, Mudlumps: Diapiric structures in Mississippi Delta sediments, in Braunstein, J., and O'Brien, G. D., eds., Diapirism and diapirs: American Association of Petroleum Geologists Memoir 8, p. 145–161.

Morgenstern, N. R., 1967, Submarine slumping and the development of turbidity currents, in Richards, A. F., ed., Marine geotechnique: Urbana, University of Illinois Press, p. 189–220.

Morris, R. H., and Pierce, K. L., 1967, Geologic map of the Vanceburg Quadrangle, Kentucky-Ohio: U.S. Geological Survey Geologic Quadrangle

GQ-2598, scale 1:24,000.

Morse, W. C., and Foerste, A. F., 1909a, The Waverlian formations of east-central Kentucky: Journal of Geology, v. 17, p. 165.

Morse, W. C., and Foerste, A. F., 1909b, The Bedford fauna at Indian Fields and Irvine, Kentucky: Ohio Naturalist, v. 9, no. 7, p. 515–523.

Morse, W. C., and Foerste, A. F., 1912, Preliminary report on the Waverlian formations of east-central Kentucky and their economic values: Kentucky Geological Survey, ser. 3, Bulletin 16, 76 p.

Myrow, P. M., and Southard, J. B., 1991, Combined-flow model for vertical stratification sequences in shallow-marine storm deposits: Journal of Sedimentary Petrology, v. 61, p. 202–210.

Nelson, C. H., 1982, Bering shelf: A mimic of Bouma sequences and turbidite systems: Journal of Sedimentary Petrology, v. 52, p. 537–545.

Nelson, C. H., Normark, W. R., Bouma, A. H., and Carlson, P. R., 1978, Thin-bedded turbidites in modern submarine canyons and fans, *in* Stanley, D. J., and Kelling, G., eds., Sedimentation in submarine fans, canyons, and trenches: Stroudsburg, Pennsylvania, Dowden, Hutchinson, and Ross, p. 177–189.

Newberry, J. S., 1870, The Geological Survey of Ohio, its progress in 1869: Report of an address delivered to the Legislature of Ohio, February 7th, 1869: Columbus, Ohio, Nevin & Myers, 66 p.

Newcombe, J. R. B., 1933, Oil and gas fields of Michigan; A discussion of depositional and structural features of the Michigan Basin: Michigan Geological Survey, Geological Series 32, Publication 38.

Nio, S.-D., 1976, Marine transgressions as a factor in the formation of sand-wave complexes: Geologie en Mijnbouw, v. 55, p. 18–40.

Nøttvedt, A., and Kreisa, R. D., 1987, Model for the combined-flow origin of hummocky cross-stratification: Geology, v. 15, p. 357–361.

O'Brien, G. D., 1968, Survey of diapirism and diapirs, *in* Braunstein, J., and O'Brien, G. D., eds., Diapirism and diapirs: American Association of Petroleum Geologists Memoir 8, p. 1–9.

Orton, E., 1879, Note on the Waverly strata of Ohio: American Journal of Science, v. 18, pt. I, p. 138–139.

Orton, E., 1888, The Geology of Ohio considered in its relations to petroleum and natural gas: Report of the Geological Survey of Ohio, v. 6, p. 1–59.

Orton, E., 1893, Geological scale and geological structure of Ohio: Report of the Geological Survey of Ohio, v. 7, pt. I, p. 3–44.

Parker, W. R., 1973, Experiments on formation of contorted structures in mud: Sedimentology, v. 20, p. 615–623.

Pashin, J. C., 1985, Paleoenvironmental analysis of the Bedford-Berea sequence, northeastern Kentucky and south-central Ohio [M.S. thesis]: Lexington, University of Kentucky, 105 p.

Pashin, J. C., 1990, Reevaluation of the Bedford-Berea sequence in Ohio and adjacent states: New perspectives on sedimentation and tectonics in foreland basins [Ph.D. thesis]: Lexington, University of Kentucky, 412 p.

Pashin, J. C., and Ettensohn, F. R., 1987, An epeiric shelf-to-basin transition: Bedford-Berea sequence, northeastern Kentucky and south-central Ohio: American Journal of Science, v. 287, p. 893–926.

Pashin, J. C., and Ettensohn, F. R., 1992a, Paleoecology and sedimentology of the dysaerobic Bedford fauna (Late Devonian), Ohio and Kentucky (USA): Palaeogeography, Palaeoclimatology, and Palaeoecology, v. 91, p. 21–34.

Pashin, J. C., and Ettensohn, F. R., 1992b, Lowstand deposition in a foreland basin: Bedford-Berea sequence (Upper Devonian), eastern Kentucky and West Virginia, *in* Ettensohn, F. R., ed., Changing interpretations of Kentucky geology—Layer-cake, facies, flexure, and eustacy: Geological Society of America Annual Meeting Field Trip Guidebook, Ohio Geological Survey Miscellaneous Report 5, p. 123–134.

Penland, S., and Boyd, R., 1985, Barrier island arcs along abandoned Mississippi River deltas: Marine Geology, v. 63, p. 197–233.

Pepper, J. F., Demarest, D. F., Merrels, C. W., II, and de Witt, W., Jr., 1946, Map of the Berea sand of southern Ohio, eastern Kentucky, and southwestern West Virginia: U.S. Geological Survey Oil and Gas Preliminary Map 69, scale 1:189,700.

Pepper, J. F., de Witt, W., Jr., and Demarest, D. F., 1954, Geology of the Bedford Shale and Berea Sandstone in the Appalachian Basin: U.S. Geological Survey Professional Paper 259, 111 p.

Posamentier, H. W., and Vail, P. R., 1988, Eustatic controls on clastic deposition; II, Sequence and systems tract models: Society of Economic Paleontologists and Mineralogists Special Publication 42, p. 125–154.

Posamentier, H. W., Allen, G. P., James, D. P., and Tesson, M., 1992, Forced regressions in a sequence stratigraphic framework: Concepts, examples, and exploration significance: American Association of Petroleum Geologists Bulletin, v. 76, p. 1687–1709.

Potter, P. E., and Pettijohn, F. J., 1977, Paleocurrents and basin analysis (2nd ed.): Berlin, Springer-Verlag, 425 p.

Potter, P. E., and Pryor, W. A., 1961, Dispersal centers of Paleozoic and later clastics of the upper Mississippi valley and adjacent areas: Geological Society of America Bulletin, v. 72, p. 1195–1250.

Potter, P. E., Maynard, J. B., and Pryor, W. A., 1980, Sedimentology of shale: New York, Springer-Verlag, 303 p.

Potter, P. E., DeReamer, J. H., Jackson, D. S., and Maynard, J. B., 1983, Lithologic and paleoenvironmental atlas of Berea Sandstone in the Appalachian basin: Appalachian Geological Society Special Publication 1, 157 p.

Pratt, T., Culotta, R., Hauser, E., Nelson, D., Brown, L., Kaufman, S., Oliver, J., and Hinze, W., 1989, Major Proterozoic basement features of the eastern midcontinent of North America revealed by recent COCORP profiling: Geology, v. 17, p. 505–509.

Prosser, C. S., 1901, The classification of the Waverly Series of central Ohio: Journal of Geology, v. 9, p. 205–231.

Prosser, C. S., 1912, The Devonian and Mississippian formations of northeastern Ohio: Ohio Geological Survey Bulletin 15, 574 p.

Quinlan, G. M., and Beaumont, C., 1984, Appalachian thrusting, lithospheric flexure, and the Paleozoic stratigraphy of the eastern interior of North America: Canadian Journal of Earth Sciences, v. 21, p. 973–996.

Rich, J. L., 1951a, Three critical environments of deposition and criteria for the recognition of rocks deposited in each of them: Geological Society of America Bulletin, v. 62, p. 481–533.

Rich, J. L., 1951b, Probable fondo origin of Marcellus–Ohio–New Albany–Chattanooga bituminous shales: American Association of Petroleum Geologists Bulletin, v. 35, p. 2017–2040.

Rittenhouse, G., 1946, Map showing distribution of several types of Berea Sand stone in West Virginia, eastern Ohio, and western Pennsylvania: U.S. Geological Survey Oil and Gas Investigations, Preliminary Map 58.

Rodgers, J., 1963, Mechanics of Appalachian foreland folding in Pennsylvania and West Virginia: American Association of Petroleum Geologists Bulletin, v. 47, p. 1527–1536.

Rodgers, J., 1967, Chronology of tectonic movements in the Appalachian region of eastern North America: American Journal of Science, v. 265, p. 408–427.

Rodgers, M. R., and Anderson, T. H., 1984, Tyrone–Mount Union cross-strike lineament of Pennsylvania: A major Paleozoic basement fracture and uplift boundary: American Association of Petroleum Geologists Bulletin, v. 68, p. 92–105.

Roep, Th. B., Beets, D. J., Dronkert, H., and Pagnier, H., 1979, A prograding coastal sequence of wave-built structures of Messinian age, Sorbas, Almeria, Spain: Sedimentary Geology, v. 22, p. 135–163.

Rominger, C. L., 1876, Geology of lower peninsula of Michigan: Michigan Geological Survey, v. 3, 161 p.

Root, S. I., and Hoskins, D. M., 1978, Lat 40°N fault zone, Pennsylvania: A new interpretation: Geology, v. 5., p. 719–723.

Root, S. I., and MacWilliams, R. H., 1986, The Suffield fault, Stark County, Ohio: Ohio Journal of Science, v. 86, p. 161–193.

Rossbach, T. J., and Dennison, J. M., 1994, Devonian strata of Catawba syncline near Salem, Virginia, *in* Schultz, A. P., and Henika, W. S., eds., Fieldguides to southern Appalachian structure, stratigraphy, and engineering geology: Blacksburg, Virginia Tech Department of Geological Sciences Guidebook 10, p. 95–125.

Rothman, E. M., 1978, The petrology of the Berea Sandstone (Early Mississippian) of south-central Ohio and a portion of northern Kentucky [M.S. thesis]: Oxford, Ohio, Miami University, 105 p.

Sanford, B. V., and Brady, W. B., 1955, Paleozoic geology of the Windsor-Sarnia area: Geological Survey of Canada Memoir 278, 65 p.

Sass, D. B., 1960, Some aspects of the paleontology, stratigraphy, and sedimentation of the Corry Sandstone of northwestern Pennsylvania: Bulletins of American Paleontology, v. 41, p. 247–381.

Schiner, G. R., and Gallaher, J. T., 1979, Geology and groundwater resources of western Crawford County, Pennsylvania: Pennsylvania Bureau of Topographic and Geologic Survey Water Resources Report 46, 103 p.

Schiner, G. R., and Kimmel, G. E., 1972, Mississippian stratigraphy of northwestern Pennsylvania: U.S. Geological Survey Bulletin 1131-A, 27 p.

Schumm, S. A., 1977, The fluvial system: New York, John Wiley, 338 p.

Schwab, W. C., and Lee, H. J., 1988, Causes of two slope-failure types in continental shelf sediment, northwestern Gulf of Alaska: Journal of Sedimentary Petrology, v. 58, p. 1–11.

Scotese, C. R., 1990, Atlas of Phanerozoic plate tectonic reconstructions: Arlington, Texas, International Lithosphere Program (IUGG-IUGS), Paleomap Project Technical Report 10-90-1, 54 p.

Sevon, W. D., 1985, Nonmarine facies of the Middle Devonian and Late Devonian Catskill coastal alluvial plain: Geological Society of America Special Paper 201, p. 79–90.

Shepard, F. P., 1955, Delta-front valleys bordering the Mississippi distributaries: Geological Society of America Bulletin, v. 66, p. 1484–1498.

Stanley, S. M., 1989, Earth and life through time (2nd ed.): New York, W. H. Freeman, 689 p.

Stow, D. A. V., and Piper, D. J. W., 1984, Deep-water fine-grained sediments: Facies and models, in Stow, D. A. V., and Piper, D. J. W., eds., Deep-water processes and facies: Geological Society of London Special Publication 15, p. 611–646.

Surlyk, F., 1987, Slope and deep shelf gully sandstones, Upper Jurassic, east Greenland: American Association of Petroleum Geologists Bulletin, v. 71, p. 464–475.

Swager, D. R., 1978, Stratigraphy of the Upper Devonian–Lower Mississippian shale sequence in eastern Kentucky outcrop belts [M.S. thesis]: Lexington, University of Kentucky, 116 p.

Swift, D. J. P., Figueiredo, A. G., Jr., Freeland, G. L., and Oertel, G. F., 1983, Hummocky cross-stratification and megaripples: A geological double standard?: Journal of Sedimentary Petrology, v. 53, p. 1295–1317.

Terwindt, J. H. J., 1981, Origin and sequences of sedimentary structures in inshore mesotidal deposits of the North Sea, in Nio, S.-D., ed., Holocene sedimentation in the North Sea basin: International Association of Sedimentologists Special Publication 5, p. 4–26.

Thomas, W. A., 1977, Evolution of Appalachian salients and recesses from reentrants and promontories in the continental margin: American Journal of Science, v. 277, p. 1233–1278.

Thomas, W. A., 1991, The Appalachian-Ouachita rifted margin of southeastern North America: Geological Society of America Bulletin, v. 103, p. 415–431.

Thompson, W. O., 1937, Original structures of beaches, bars, and dunes: Geological Society of America Bulletin, v. 48, p. 723–752.

Tomlinson, C. W., 1916, The origin of red beds: American Journal of Science, v. 24, p. 153–179.

van den Berg, J. H., 1977, Morphodynamic development and preservation of sedimentary structures in two prograding recent ridge and runnel beaches along the Dutch coast: Geologie en Mijnbouw, v. 56, p. 185–202.

van Houten, F. B., 1973, Origin of red beds: A review—1961–1972: Annual Review of Earth and Planetary Sciences, v. 1, p. 39–42.

van Siclen, D. C., 1958, Depositional topography—examples and theory: American Association of Petroleum Geologists Bulletin, v. 42, p. 1897–1913.

van Steenwinkel, M., 1990, Sequence stratigraphy from "spot" outcrops: Example from a carbonate-dominated setting: Devonian-Carboniferous transition, Dinant synclinorium (Belgium): Sedimentary Geology, v. 69, p. 259–280.

van Steenwinkel, M., 1993, The Devonian-Carboniferous boundary in southern Belgium: Biostratigraphic identification criteria of sequence boundaries, in Posamentier, H. W., Summerhayes, C. P., Haq, B. U., and Allen, G. P., eds., Sequence stratigraphy and facies associations: International Association of Sedimentologists Special Publication 18, p. 237–246.

Vause, J. E., 1959, Underwater geology and analysis of Recent sediments off the northwest Florida coast: Journal of Sedimentary Petrology, v. 29, p. 555–563.

Vavrdová, M., Isaacson, P. E., Díaz, E., and Beck, J., 1991, Palinologia del limit Devonico-Carbonifero entorno al Lago Titikaka, Bolivia: Resultados preliminares: Revista Technica, Yacimentos Petroliferos Fiscales Bolivianos, v. 12, p. 303–313.

Wagner, W. R., 1976, Growth faults in Cambrian and Lower Ordovician rocks of western Pennsylvania: American Association of Petroleum Geologists Bulletin, v. 60, p. 414–427.

Walker, R. G., 1978, Deep-water sandstone facies and ancient sub-marine fans: Models for exploration and stratigraphic traps: American Association of Petroleum Geologists Bulletin, v. 62, p. 932–966.

Walls, R. A., 1975, Late Devonian–Early Mississippian subaqueous deltaic facies in a portion of the southeastern Appalachian basin, in Broussard, M. L., ed., Deltas: Houston, Texas, Houston Geological Society, p. 359–367.

Wanless, H. R., and 9 others, 1970, Late Paleozoic deltas in the central and eastern United States: Society of Economic Paleontologists and Mineralogists Special Publication 15, p. 215–245.

Waschbusch, P. J., and Royden, L. H., 1992, Episodicity in foredeep basins: Geology, v. 20, p. 915–918.

Wasson, T., and Wasson, I., 1929, Cabin Creek field, West Virginia, in Structure of typical American oil fields: A symposium: Tulsa, American Association of Petroleum Geologists, p. 462–475.

Wells, N. A., Coogan, A. H., and Majoras, J. J., 1991, Field guide to Berea Sandstone outcrops in the Black River Valley at Elyria, Ohio: Slumps, slides, mud diapirs, and associated fracturing in Mississippian delta deposits: Ohio Journal of Science, v. 91, p. 35–48.

White, I. C., 1880, The geology of Mercer County Pennsylvania: Second Geological Survey of Pennsylvania Report QQQ, 233 p.

White, I. C., 1881, The geology of Erie and Crawford Counties: Second Geological Survey of Pennsylvania Report QQQQ, 406 p.

Whittlesey, C., 1838, Mr. Whittlesey's Report: Columbus, First Annual Report of the Geological Survey of Ohio, p. 41–71.

Woodrow, D. L., and Isley, A. M., 1983, Facies, topography, and sedimentary processes in the Catskill Sea (Devonian), New York and Pennsylvania: Geological Society of America Bulletin, v. 94, p. 459–470.

Wright, L. D., Chappell, J., and Thom, B. G., 1979, Morphodynamics of reflective and dissipative beach and inshore systems, southeastern Australia: Marine Geology, v. 32, p. 105–140.

Ziegler, A. M., and McKerrow, W. S., 1975, Silurian marine red beds: American Journal of Science, v. 275, p. 31–56

MANUSCRIPT ACCEPTED BY THE SOCIETY AUGUST 5, 1994

Typeset and printed in U.S.A. by Johnson Printing, Boulder, Colorado